T0263081

TROPICAL MEDICINE: AN ILLUSTRATED HISTORY OF THE PIONEERS

TROPICAL MEDICINE: AN ILLUSTRATED HISTORY OF THE PIONEERS

G C Cook MD, DSC, FRCP, FRCPE, FRACP, FLS

Visiting Professor, University College, London, UK

ELSEVIER

PARIS • AMSTERDAM • BOSTON • HEIDELBERG • LONDON • NEW YORK • OXFORD
PARIS • SAN DIEGO • SAN FRANCISCO • SINGAPORE • SYDNEY • TOKYO

Academic Press is an imprint of Elsevier

Academic Press is an imprint of Elsevier
84 Theobald's Road, London WC1X 8RR, UK
360 Park Avenue South, New York, NY 10010-1710
30 Corporate Drive, Suite 400, Burlington, MA 01803, USA
525 B Street, Suite 1900, San Diego, California 92101-4495, USA

First edition 2007

British Library Cataloguing in Publication Data
A catalogue record for this book is available from the British Library

Library of Congress Cataloging in Publication Data
A catalog record for this book is available from the Library of Congress

ISBN: 978-0-12-373991-9

For information on all Academic Press publications
visit our web site at books.elsevier.com

Typeset by Charon Tec Ltd (A Macmillan Company), Chennai, India
www.charontec.com

Working together to grow
libraries in developing countries

www.elsevier.com | www.bookaid.org | www.sabre.org

ELSEVIER   BOOK AID
International   Sabre Foundation

CONTENTS

PREFACE

This book is primarily an account of the individuals who made major contributions to, and indeed shaped, the formal discipline of tropical medicine in the late nineteenth and early twentieth centuries; in fact in the latter years of Victoria's reign (see Figure A).[1] The origin(s) of this discipline (dominated by a handful of 'prima donnas') is very largely a result of much foresighted action by the Seamen's Hospital Society, coupled with the enthusiasm of several politicians of a bygone era.

As recently as 2001, however, a Harveian Orator at the Royal College of Physicians spoke as though this discipline and 'medicine in the tropics' were one and the same. He concentrated on: malaria, snake bite and rabies, and concluded:[2]

> In the 350 years of its history, this is the first Harveian Oration to be devoted to Tropical Medicine; I hope that this is a sign that my speciality has at last [now that the formal discipline is in steep decline] been accepted into the mainstream of medicine in this country.

Those who have worked for long in warm climates will, however, recognize that 'medicine in the tropics' has existed since time immemorial, and that it should be clearly demarcated from the rise and fall of the *formal* discipline.

FIGURE A   Queen Victoria (1819–1901) (reproduced courtesy of The Wellcome Library, London).

Lloyd and Coulter, in their classic text *Medicine and the Navy 1200–1900*, have admirably summarized the difference between the two:[3]

> For the historian of medicine, the fact that Naval Surgeons [before the development of the formal specialty] had more experience of *tropical diseases* [my italics] than any other branch of the profession … as Sir Patrick Manson recognized when he founded that discipline in England. It cannot be pretended that they made much use of their experience because, for the most part, they were not outstanding members of their profession. None the less, they were by no means more backward than their colleagues on land before the days when the germ theory of disease replaced the atmospheric or climatorial pathology which prevailed up to the last decade of the [nineteenth] century. Whatever new advances were made in medical practice (and they were not numerous before the days of Lister and Pasteur) they were adopted in the [Naval] service with commendable promptness.

Thus, experience in the Caribbean, Africa and the 'jewel in the Crown' – India[4] – pre-dated the development of the formal discipline, as did the practice of medicine in the West African Squadron. But even before this, much information on the diagnosis and management of disease acquired in tropical countries had been gathered over many centuries. Andrew Balfour (1873–1931)[5] recognized this when he recalled, in his Presidential Address to the Royal Society of Tropical Medicine and Hygiene in 1925:[6]

> There is in one sense no such thing as tropical medicine … many of the most erudite writings of Hippocrates are concerned with maladies which now-a-days are chiefly encountered under tropical or sub-tropical conditions.

An anonymous contributor to the *British Medical Journal* for 1913, summarizing the position a century before this, wrote:[7]

> The study of *tropical diseases* [my italics] was obviously in its infancy in 1813; but there were indications that British medical men in India and in the West Indies were beginning to turn their attention to it, being forced to it indeed by the mortality from such maladies which was making itself felt in the troops stationed abroad and among the sailors manning the Indiamen and the fighting frigates of the time. There were no such schools of tropical medicine in those days such as are to be seen now; but there was already a clear indication of the necessity which was to be laid upon Great Britain of the later nineteenth and of the earlier twentieth centuries to send forth of the best of her medical sons to discover the cause of Malta fever, to unravel the problem of sleeping sickness, and to learn to slay many another scourge existing in these tropical countries over which she has been placed as ruler, and for which she has already paid a price in blood.

Thus, before development of the formal discipline, much of the practice of 'medicine in the tropics' fell under the heading of hygiene, defined by the *Oxford English Dictionary* as:[8]

> That department of knowledge or practice which relates to the maintenance of health; a system of principles or rules for preserving or promoting health; sanitary science.

The development of the 'formal discipline', which focused largely on teaching the medical officers of numerous British colonies situated in warm climates about the 'exotic' diseases which the Western practitioner knew little or nothing about, has with some justification been branded *colonial medicine* by some medical historians; Patrick Manson (1844–1922; see Chapter 3) was clearly the founding father of the discipline. According to G C Low (1872–1952; Chapter 8), an important but grossly under-recognized pioneer of the discipline, it was established during a 20-year period, 1894–1914.[9]

It is also crucially important to appreciate that 'medicine in the tropics' encompasses not only the 'exotic' infections (which have enormous geographical variations, e.g. many South American diseases simply do not exist in Africa), most of them parasitic in origin, but also more mundane infections and diseases which even today do not fall within the 'formal discipline' but occupy most of the time of the medical practitioner serving in the tropics – HIV/AIDS,[10] tuberculosis, pneumococcal phemomia and chronic rheumatic cardiac disease, for example, constitute the 'bread and butter' of medical practice there. Furthermore, populations in the tropics are rapidly becoming more urbanized, as happened in Britain in the early to mid-nineteenth century; therefore understanding of 'medicine in the tropics' must keep up with the times.[11] That there is significant overlap between these two entities cannot, however, be denied.

The early years of the 'formal discipline' were closely associated with the activities of the (Royal) Society of Tropical Medicine and Hygiene, which was founded in the early twentieth century and provided a forum where many of the enigmas relating to the exotic infections, most of parasitic origin, were discussed by the pioneers.

Modern historians will doubtless criticize me for not including any women, and furthermore for concentrating on British contributions. However, none of the pioneers of the specialty was a woman, and the vast majority of the contributions of significance were made by British workers. In view of the fact that Britain, with her vast Empire, led the way, it is not surprising that the majority of the pioneers were British.[12] Even revisionist theories – so favoured by present-day historians – must be based on fact, otherwise, history will become no more than fiction!

Many previous works have addressed the subject from the point of view of disease entities (see, for example, *The Wellcome Trust Illustrated History of Tropical Diseases*[13]), or have viewed the discipline from a somewhat narrow perspective. I have therefore attempted to orient this book around the pioneers themselves.

G C Cook
December 2006

NOTES

1 G C Cook. *From the Greenwich Hulks to Old St Pancras: a history of tropical disease in London.* London: Athlone Press, p. 338; G C Cook, A Zumla (eds) (2003), *Manson's Tropical Diseases*, 21st edn. London, 1992: WB Saunders, p. 1847.

2 D Warrell. *'To search and studdy out the secrett of tropical diseases by way of experiment': The Harveian Oration.* London, 2001: Royal College of Physicians, p. 41.

3 C Lloyd, J L S Coulter. *Medicine and the Navy 1200–1900*, Vol. 4, 1815–1900. London, 1963: E&S Livingstone, p. vi.

4 D G Crawford. *A History of the Indian Medical Service 1600–1913*, Vols I and II. London, 1914: W Thacker & Co., pp. 529, 535.

5 Sir Andrew Balfour was the first (and only) Director of the London School of Hygiene and Tropical Medicine. See also A S MacNalty, M E Gibson. Balfour, Sir Andrew (1873–1931). In: H C G Matthew, B Harrison (eds), *Oxford Dictionary of National Biography*, Vol. 3. Oxford, 2004: Oxford University Press, pp. 493–4.

6 A Balfour. Some British and American pioneers in tropical medicine and hygiene. *Trans R Soc Trop Med Hyg* 1925, 19: 189–231.

7 Anonymous. One hundred years ago. *Br Med J* 1913, i: 455–6.

8 J A Simpson, E S C Weiner (eds). *The Oxford English Dictionary*, 2nd edn, Vol. 7. Oxford, 1989: Clarendon Press, p. 546.

9 G C Low. A retrospect of tropical medicine from 1894–1914. *Trans R Soc Trop Med Hyg* 1929, 23: 213–34.

10 E Moore. HIV is changing the face of tropical medicine. *Br Med J* 2006, 332: 1280. See also C Apetrei, P A Marx, S M Smith. The evolution of HIV and its consequences. *Infect Dis Clin N Am* 2004, 18: 369–94.

11 G C Cook. Tropical medicine as a formal discipline is dead and should be buried. *Trans R Soc Trop Med Hyg* 1997, 91: 372–4. See also G C Cook. Future of tropical medicine. *Br Med J* 1996, 312: 1160.

12 R Desmond. *Victorian India in Focus: a selection of early photographs from the collection in the India Office Library and Records*. London, 1982: HMSO, p. 100; J. Morris. *Heaven's Command: An Imperial Progress*. London, 1992: The Folio Society, p. 470; J. Morris. *Pax Britannica: The Climax of an Empire*. London, 1992: The Folio Society, p. 408; P J Marshall (ed.). *The Cambridge Illustrated History of the British Empire*. Cambridge, 1996: Cambridge University Press, p. 400.

13 F E G Cox (ed.). *The Wellcome Trust Illustrated History of Tropical Diseases*. London, 1996: The Wellcome Trust, p. 452.

PROLOGUE

What denotes a pioneer? The *Oxford English Dictionary* definition is:[1]

> One who goes before to prepare or open up the way for others to follow; one who begins, or takes part in beginning, some enterprise, course of action, etc; *an original investigator, explorer, or worker* [my italics], in any department of knowledge or activity; an originator, initiator (*of* some action, scheme, etc); a forerunner (in such action, etc).

Clearly, therefore, pioneers of tropical medicine originate from the first occasion on which *Homo sapiens* visualized a macroparasite, e.g. *Ascaris lumbricoides* (the round worm) or *Dracunculus medinenisis* (the guinea worm). I shall thus attempt to limit the use of *pioneer* to one who made a specific contribution to the cause, prevention or management of one (or more) of the diseases which is more common in a warm climate (usually associated with substandard hygienic conditions) than in a temperate one.[2]

In Chapter 1, I have included some of those who made pertinent observations regarding 'disease in the tropics', which opened the way for others to follow, but did not necessarily clinch advances made by scientific verification. While the major pioneers of the formal discipline are relatively easy to identify, this certainly does not apply to their predecessors; I have again given examples of the latter in Chapter 1. Prior to the enunciation of the 'germ theory', it is virtually impossible to accredit a discovery to a given individual.

The great 'explosion' of major advances, which gave birth to the formal discipline, resulted indirectly from the demonstration that all disease entities have specific causes (Chapter 2). This was, of course, dependent on acceptance of the 'germ theory' of disease – as outlined by Pasteur and, later, by Koch and Lister.

Although some have made important literary contributions to the subject, this does not necessarily allow them to be designated pioneers. For example, numerous British publications in the seventeenth century praised the curative properties of the *bark* in intermittent fevers, but the individual who first demonstrated the febrifuge properties of cinchona will almost certainly never be identified.

Many of the pioneers of the formal discipline (which was dominated by clinical parasitology) became Presidents of the (Royal) Society of Tropical Medicine and Hygiene, and have left addresses, often outlining their contribution(s), in the *Transactions* of that Society. This Society was therefore closely linked with the major pioneers of the discipline.[3]

Early history of the formal discipline of tropical medicine incorporates various contentious issues, many of them surrounding malaria research. For example: should Manson have received a Nobel Prize? He was probably the first to implicate the mosquito in human infection scientifically, and was Ross's mentor throughout his seminal malaria work in India. Also, should the Italian malariologists have received greater recognition in the solution of the mosquito–man cycle? I shall address these issues and many more in these pages; however, the final adjudication must rest with the reader.

NOTES

1   J A Simpson, E S C Weiner (eds). *The Oxford English Dictionary*, 2nd edn, Vol. 11. Oxford, 1989: Clarendon Press, p. 883.

2   G C Cook. History of parasitology. In: S H Gillespie, R D Pearson (eds), *Principles and Practice of Clinical Parasitology*. Chichester, 2001: John Wiley & Sons, pp. 1–20; F E G Cox. History of human parasitic diseases. *Infect Dis Clin North Am* 2004, 18: 171–88.

3   G C Cook. Evolution: the art of survival. *Trans R Soc Trop Med Hyg* 1994, 88: 4–18.

1

Early pioneers of 'medicine in the tropics'

Until the 'germ theory' of disease (see Chapter 2) was widely accepted, in the latter part of the nineteenth and the early twentieth centuries, there was no universal acceptance of disease causation. The miasmatists held that disease lurked in the ground and thus had a 'telluric' origin;[1] the contagionists maintained that diseases were conveyed from one individual to another.[2]

THE BOTANISTS AND CLIMATOLOGISTS

Many early encounters with tropical diseases took place in the navy, where the most common disease entities were 'typhus' and scurvy. Tropical diseases were experienced to a great extent by the West African Squadron; in attempts to prevent the West African slave trade from continuing, their ships frequently penetrated inland waterways.[3] Early explorers concentrated on climate and geology as being important in the *causation*, and botanical factors in the *cure*, of these 'exotic' diseases. The *effective* pharmaceutical list was, in fact, exceedingly limited (see Table 1.1) and generally ineffective. The 'fever', much of which must have been caused by malaria, was of course a dominant theme.[4] This was the case in the explorations by David Livingstone (1813–73; Figure 1.1) in south, central and east Africa.[5]

TABLE 1.1 The pharmaceutical armamentum of the traveller to the tropics in 1832*

Drug	Quantity	Description
Submuriate of Mercury of Calomel	1 lb	Purgative
Colocynth	2 lb	–
Epsom Salts	10 lb	–
Jalap	1 lb	–
Seydlitz Powders	12 doz.	–
Tartar Emetic	1 oz	Emetic
Ipecacuanha	4 oz	–
James's Powders	6 pkts	Sudorific
Citric Acid	2 lb	For lemon juice
Carbonate of Soda	2 lb	For saline drink
Dover's Powders	1 lb	Sudorific
Spirits of Nitre	1 lb	–
Cream of Tartar	4 lb	For acidulated drink
Opium	4 lb	Anodyne
Vitriolic Ether	8 oz	Stimulant
Spirit of Hartshorn	1 lb	–
Camphor	4 oz	–
Blue Pills (mercurial)	8 oz	Laxative
Sulphate of Quinine	4 oz	Strengthener after fever or dysentery
Blistering powder, bandages, lancets, sponges, etc.		

*See R Lander (1832) *Journal of Expedition to Niger*, Vol. 3, p. 348.

Most of this generation of physicians/surgeons were thus essentially medical botanists, whose major ambition was to assemble large collections of medical plants for inclusion in the British Museum – such as that built up by Sir Joseph Banks (Figure 1.2), President of both the Royal Society and the Royal College of Physicians. Other collections were housed at the Natural History Museum, London; the Royal Botanical Gardens at Kew and Edinburgh; the Liverpool Museum; and the Linnean Society of London.

EARLY ACCOUNTS OF 'MEDICINE IN THE TROPICS'

The distinguished medical historian Charles Singer (1876–1960) has brought to the fore some of the early references to tropical diseases. Beginning with the

$\mathcal{David\ Livingstone}$

FIGURE 1.1   David Livingstone (1813–73) provided the first literary contributions to medicine in Africa (reproduced courtesy of The Wellcome Library, London).

reign of Elizabeth I (1533–1603), in which new and strange lands were discovered by mariners from the Western world, he documents what is probably the earliest British work intended for sailors: *The CVRES of the Diseased, in Remote Regions, Preventing Mortalitie, incident in Forraine Attempts, of the English Nation. La honra mas vale merecerla que teneria*, by 'GW' (1598). The author

FIGURE 1.2   Joseph Banks FRS (1743–1820) was an early medical botanist (reproduced courtesy of The Wellcome Library, London).

had voyaged mostly to the Spanish Main, and he began his book with these lines in verse:

> The burning fever, calde the *Calenture*
> The aking *Tabardilla* pestilent,
> Th' *Espinlas* prickings which men do endure,
> *Cameras de Sangre*, Fluxes violent
> Th' *Erizpila* swelling the Pacient,
> The *Tinoso*, which we all Scurvey call
> are truly here describ'd and cured all.

Calenture, according to Singer, probably refers to malaria, *Tabardilla* to yellow fever (or possibly a form of malaria, dengue, or typhus fever), and *Cameras de Sangre* to the bloody flux; the disease to which *Espinlas* refers has not yet been identified.

Singer was of the opinion that *Trypanosoma gambiense* infection was first clearly documented by a naval surgeon, John Atkins (1685–1757),[6] in *The Navy*

Surgeon ... (1721). It had previously been felt that the description by Thomas Winterbottom (1766–1859) in Sierra Leone in 1803 was the first. Atkins considered that the cause was

> a superabundance of phlegm or serum extravasated in the brain, which obstructs the irradiation of the nerves, [and the treatment was] ... to rouse the spirits by bleeding in the jugular, quick purges, Sternatories, Vesicatories, Acupuncture, Seton, Fontanels, and Sudden Plunges in the Sea; the latter [being] most effectual when the Distemper is new and the Patient as not yet attended with a drivling at Mouth and Nose.

According to Singer, the sand flea (*Pulex penetrans*) was first documented in *Cronica de las Indias* (1547) by Fernandez de Oviedo.

The macro parasite *Dracunculus medinensis* was well known to the ancient world, and Singer has outlined its history. Understanding of this helminthiasis has changed little with the passage of time. Reference to it was made by Plutarch, Galen, Aetius and Paulus Aegineta; some apparently held that the fiery serpents which tormented the children of Israel in the desert were examples of *D. medinensis*; the Arabian physicians gave full accounts; and in the sixteenth century the Portuguese Jew Amatus Lusitanus (physician to Pope Julian III) wrote in Salonica on the 'Turkish disease':

> a dangerous malady which develops not only in the country from which it takes its name, but also in Egypt, India, and other countries as the Arabian physicians, especially Avicenna and Avenzoar, do teach, describing it as the Medina vein.

Regarding treatment of dracontiasis, Amatus quotes a physician who was learned in Arabic:

> First, the patient ties the end of the vein or nerve [there was apparently uncertainty as to whether it was a nerve, vein or helminth] round a small piece of wood, and this he winds little by little till the last part of the worm is drawn out. As the structure is often three cubits long, the treatment may last many days before the sufferer is altogether free from pain and inconvenience. Many adopt a cataplasm or cold suffusion, as Soranus, Leonides, and Paulus of Aegina recommend.

The Dutch explorer, Linschoten, also described *D. medinensis* in 1599, and provided a sketch of the helminth. The discovery that infection was the result of a contaminated water supply is attributed to a former Bishop of Beirut who, in the mid-seventeenth century, considered:

> to avoid this worm is to drink only wine, or if water is used, only such as has been carefully filtered through linen.

Singer continued:

> The very emblem of our profession [gave G H Velschius the idea of depicting] serpents coiling round the staff of Aescaulepius [illustrating] a guinea-worm which [he] has just extracted by entwining it around a piece of wood.

The history of yaws was also researched by Singer. The Spanish physician Oviedo (see above) gave it the name 'bubas', and described it in *Natural Hystoria de las Indias* (1526) and *Cronica de las Indias* (1547). It was termed 'pians' by André Thevet in *France Antarctique* (1558). There was a great deal of confusion with syphilis, as well as regarding its country of origin. Oviedo wrote:

> I was often amused in Italy by hearing the Italians speak of the 'French disease', whilst the French call it the 'disease of Naples'. But of a truth they would better hit its name off if they called it the disease of the Indies. And that this is the truth may be gathered from this chapter and from many experiments already made with holy wood and guayacan, where-with especially, better than any other medicine, this terrible pustulous disease is cured and healed.

The use of guaiacum as a *native* treatment of yaws, wrote Singer, survived its use in syphilis!

George Playfair (1782–1846)[7] was in 1813 serving as an Assistant Surgeon in Bengal. He used ipecacuan (30–60 grains) + laudanum (a similar amount) for dysentery. Disclaiming all originality, he wrote:

> I recollected to have seen it in, I think, a Medical Review, that ipecacuan and opium had been tried in large doses with effect.

Playfair had apparently used this combination when he was a surgeon of an Indiaman in 1803–4. Much later Rogers (see Chapter 13) also recommended the use of ipecacuanha, and had (in 1912) used emetine (the principal alkaloid of ipe-cacuanha) by 'hypodermic' injection. It had, according to him, a specific effect on amoebic dysentery. Also addressed was the management of rabies by 'bleeding'. Fothergill had used this remedy in 1774, but only to the extent of 6 oz, which was insufficient for a cure. The article emphasized that 'bleeding in hydropho-bia is now [in 1913] only of historical interest'. Furthermore that Louis Pasteur's (1822–95; see Chapter 2) method of prevention (by inoculation in the incubation period) had replaced all other plans. Yellow (tropical or Caribbean) fever was also treated by 'bleeding'.[8]

These articles provide a reminder that, in 1812–13, British ships were cruis-ing in West Indian waters, for Britain was at war with America and, furthermore, the ships also had 'to keep watch against French privateers in the Caribbean sea'; and 'naval surgeons and physicians were at that time building a knowledge of tropical medicine'. Thus, clinical aspects of many of the diseases existing in the torrid zone were clearly documented long before the 'germ theory' of disease had been formulated. However, it is exceedingly difficult, if not impossible, to identify a true *pioneer* for any one of them.

Several so-called tropical diseases were of course prevalent in England until the twentieth century. Harold Scott (1874–1956) stressed that

> agues were common [in the mid nineteenth century] in the Fens [and] in the marshes of the Thames Estuary, in Kent, Essex, and south coasts, on the Romney and Pevensey marshes, [and] at Bridgwater near the Bristol Channel.

The building of the Thames Embankment by Joseph Bazalgette (1819–91) rid London of *Plasmodium vivax* infection at last.[9]

FEVERS IN HISTORY

In fact, the close association between swamps and marshes and the prevalence of 'agues' and 'marsh fevers' had been common knowledge for many centuries; as Scott stressed, however, there were also extensive marshes where neither of these illnesses existed – a fact that perplexed many observers. Although David Livingstone recorded (like earlier observers) that mosquitoes were a menace, a direct link with 'ague' was not made; the supposition continued that 'the fever' resulted from miasmas and (decaying) vegetative ferments, and was caused by 'the malaria poison'.

The term that has been used since the eighteenth century for this disease – *malaria* (bad air) – is, of course, now known to be totally erroneous! 'Paludism' is also a misnomer, for, as Scott has emphasized (see above), 'there are [for example] marshes without malaria and in a great many places malaria without marshes'. There have, of course, been numerous suggestions explaining the link with marshes.

As long ago as 36 BC Vorro, and in AD 100 Columella, suggested that the 'fever' was caused by organisms too small for the eye to see, and the latter also noted that the bite of the mosquito might transmit these organisms. Hensinger suggested in 1844 that the 'miasma' might be a poison or parasite. There was also a suggestion that the miasmas might be driven by wind.[10] A theory that drinking water caused the disease had existed for at least two centuries, and was still considered likely in the very late nineteenth century; coupled with that was a suggestion that, after heavy rains, organisms are 'washed into … shallow wells, without undergoing filtration through … deeper strata'. Many different measures have been tried throughout the ages to encounter the 'marsh fevers', which undoubtedly included not only malaria (see Chapter 4) but also numerous other febrile diseases. The value of mosquito-nets in prevention of 'the fever' has been well known since Roman times.[11]

'Fevers' in the eighteenth and nineteenth centuries

As William Bynum (1943–present) has pointed out, the number of works on this subject deemed worthy of inclusion in Garrison and Morton's bibliography of landmarks in the history of medicine, despite numerous 'competitors', is very limited. Nosology in the eighteenth century was entirely symptom-based. Fevers (which were *not*, as is the case today, equated simply with raised body temperature) of all diseases, were most obviously related to geography and climate. With the expansion of the British Empire and the establishment of trading routes to India, the East and West Indies, and parts of the African (especially West) coast, together with the supporting military, naval and civilian establishments, this was foremost in the minds of the physicians over several hundred years before the 'germ theory' became dominant. It is, as Bynum has again pointed out, striking how much of the 'fever literature' of the second half of the eighteenth century was produced by men (in most cases with a Scottish background)

JAMES LIND, M.D.

FIGURE 1.3   James Lind (1716–94) made contributions to medicine in the British Navy (reproduced courtesy of The Wellcome Library, London).

who had practical experience abroad, many of whom initiated debates concerning the nature and treatment of fevers which included not only those indigenous to Britain. James Lind (1716–94; Figure 1.3),[12] John Pringle (1707–82; Figure 1.4),[13] William Cullen (1710–90),[14] John Coakley Lettsom (1744–1815; Figure 1.5)[15] and Robert Robertson (1742–1829; Figure 1.6)[16] all fell into this category. Pelling has rightly stressed that it was in fact 'fever' and not cholera, which was the *driving force* behind the public health policies of Victorian Britain.

Lind's *Essay on Diseases Incidental to Europeans in Hot Climates* (1768) attempted to relate fevers, dysenteries, and other acute diseases prevalent in India, Africa and the West Indies to the environmental milieu pertaining there; ambient temperature and other characteristics of the air seemed pertinent to the causation of these diseases. Fresh air could be protective – not only in the macroscopic case of miasmatic marsh vapours, but also in the instance of contagion

FIGURE 1.4   John Pringle (1707–82) made contributions to medicine in the British Army (reproduced courtesy of The Wellcome Library, London).

spread from person-to-person. The views of Lind and Pringle (and also Rush) on cleanliness were of course reinforced by John Wesley (1703–91). The prevailing view (challenged by Lind) was that many tropical diseases were qualitatively different from those found in Europe, and that a white traveller was unlikely to adapt to them.

Robertson, like most contemporaries, considered that all fever was identical and 'more or less infectious' – that there was in fact a unity of all fevers – and, furthermore, 'febrile infection has been, in all ages and in all situations, one and the same idiopathic disease'. But he also stressed, as had Lind before him, that there was a host element: 'the subsequent phenomena of the fever will depend chiefly on the state of the patient's system when he is infected'. Much of the 'fever' he encountered must have been caused by malaria or yellow fever; Robertson also had a great deal of experience of typhus (see below). After more than 30 years in the Royal Navy (during which he had served for a considerable time on the

FIGURE 1.5   John Coakley Lettsom (1744–1815) made early contributions to preventive medicine (reproduced courtesy of The Wellcome Library, London).

West coast of Africa and the West Indies) he reached this conclusion. In his opinion also, Peruvian bark (see below) was indicated in all and efficacious in some of them. Bleeding, which was currently in widespread use (and was advocated by authorities such as John Huxham, 1692–1768), was, however, of no avail; he considered that bleeding reduced the heat and velocity of the blood, and allowed absorption of diluting liquids into the bloodstream. Overall, he felt that, used early and appropriately, the bark would cure fever in all climates and conditions.

George Roupell (1797–1854)[17], a physician working on the first and second of the Seamen's Hospital Society's hospital ships at Greenwich, gave a breakdown of diagnoses over a three-year period. In his report for 1829 he separated 'continuous' from 'intermittent' fevers; of 293 cases, 112 were 'intermittent' and 181 'continuous' (see Table 1.2).

However, there were, as today, unorthodox practitioners, one of whom proved (though not in his lifetime) to be correct. Francis Riollay (*c*. 1748–1797)[18] recognized the disparity between theory and practice in the late eighteenth-century understanding of fever. He insisted that

ROB^T ROBERTSON, M.D. F.R.S. F.A.S.

Member of the Royal College of Physicians
and Physician to the Royal Hospital at Greenwich.

FIGURE 1.6 Robert Robertson (1742–1829) made early recordings of disease prevalence in the tropics (reproduced courtesy of The Wellcome Library, London).

> *fever* is not a *disease* in itself; in all cases, it is *symptomatic* of some affection; and never *primary* or *idiopathic*. It is more natural to think that fever is a symptom of a *particular* affection; it is also a symptom when the affection is *general*. Fever seems in fact to be *Nature's 'common signal of distress'*.

'Fevers' in the late nineteenth century, before Laveran and Ross

(See also Chapters 4 and 5.)

This is not the place to attempt to trace the history of malaria, but suffice it to say that there is probable evidence from several continents that tertian and quartan fevers, and splenomegaly, were well documented several thousand years ago. The theory of a causative role of a poisonous vapour (miasma) dates back to at least the ancient Greeks. Effluvia, or exhalations from the marshes (vegetable decomposition) or from within the Earth, caused fevers. Fevers – which were periodic, and associated with climate, season and geographical location – were

TABLE 1.2 Medical cases admitted to the Seamen's Hospital Society's ships HMS *Grampus* or HMS *Dreadnought* 1829–31; deaths in parentheses*

Date (year)	1829	1830	1831
Acute			
Fevers	293 (7)	273 (6)	387 (33)
Chest diseases	118 (10)	126 (15)	108 (12)
Abdominal diseases	62 (10)	69 (5)	84 (13)
'Brain' diseases	28 (3)	24 (7)	9 (2)
Rheumatism	91 (0)	141 (2)	125 (2)
Chronic			
Phthisis	30 (25)	22 (12)	30 (14)
Heart disease and dropsy	8 (2)	25 (13)	30 (11)
Scurvy	19 (0)	11 (0)	10 (0)
Stomach diseases	29 (0)	–	24 (0)
Others	–	10 (1)	14 (1)

* From data compiled by G L Roupell (1797–1854) in the Seamen's Hospital Society archive. See note 17 (Cook, 2005).

frequently referred to as 'paludal' (see above). It was well known that such fevers were *not* contagious.

In 1811, James Boyle (dates unknown)[20] had given a great deal of support for the swamp theory. It was also well known that long-term residents in a malarious area (such as the Roman Campagna) developed relative resistance to the disease. Indeed, the idea that drinking contaminated marsh water caused this disease was accepted as recently as the last decade of the nineteenth century – prior to Ross's confirmation of the role of the mosquito in transmission. Winds that came from marshes were also felt to be potentially dangerous. The 'miasma theory' of disease in fact persisted after the true cause of malaria had been delineated (see Chapters 4 and 5).

Is an infective agent involved in the intermittent fever?

As long ago as 1716, Giovanni Lancisi (1654–1720; Figure 1.7) noted that human brains and spleens in individuals dying of fever contained black pigment. In 1876, Joseph Jones (1833–96),[21] working in New Orleans, wrote in his memoirs:

> Many of the colorless [red] corpuscles also contain numerous dark granules, as if they had fed upon the altered blood corpuscles and appropriated them [and] Globular bodies were found not only in the *liquor sanguinis* but also in the colored corpuscles and in the colorless corpuscles. These bodies are most probably true spores and appear to possess the power of invading and destroying the colored corpuscles.

Laveran (see Chapter 4) was of course the investigator who first recognized that the intra-erythrocytic 'spores' were in fact living organisms, and thus the

FIGURE 1.7   Giovanni Lancisi (1654–1720) provided the first documentation of malarial pigment in *Homo sapiens* (reproduced courtesy of The Wellcome Library, London).

centuries-old 'miasma' theory came to a gradual end, even though Hieronymous (Girolamo) Fracastorius (1478–1553) had outlined the 'animalcular theory' as early as 1546. As Poser and Bruyn have outlined in their erudite account of malaria, many had in fact dropped the miasmatic and supported the animalcular origin of malaria long before Laveran's demonstration in 1880 (see Chapter 4).[22]

GEOGRAPHICAL CONTRIBUTIONS TO DISEASE PREVALENCE

Many of the pioneers of 'disease in the tropics' wrote important treatises on disease nature and prevalence in specific geographical locations. Unless something was known of the presence (and epidemiology) of disease in the immediate locality, there was little that could be done regarding dealing with it! Therefore, many pioneers before the 'germ theory' developed botanical cures for the local diseases. A book written by William Twining (1790–1844) in Bengal is exemplary in this respect.[23]

Before comprehensive books devoted to tropical diseases (such as *Manson's Tropical Diseases*, 1898) were written, the practitioner was heavily dependent on local descriptions of disease. These descriptions were made wherever the British Empire was influential, and usually centred on the health of the expatriate rather than the indigenous population.

The Caribbean

Many of the earliest treatises on 'medicine in the tropics' emanated from the West Indies. Although Hans Sloane (1660–1753; Figure 1.8)[24] spent 15 months in Jamaica as physician to the newly appointed Governor, the second Duke of

FIGURE 1.8 Hans Sloane (1660–1753) researched medical botany in the Caribbean (reproduced courtesy of The Wellcome Library, London).

Albemarle, his research contribution was largely limited to medical botany. He had already been elected a Fellow of the Royal College of Physicians (of which he later became President, on 12 April 1687) when he reached Jamaica on 19 December 1657, aged 34 years.

On returning to England from Barbados, William Hillary (1697–1763) wrote, in 1759, one of the first books in English solely devoted to medicine in a tropical location: *Observations on the Changes of the Air, and the concominant epidemical disease, in the Island of Barbados, to which is added a treatise on the putrid bilious fever, commonly called the yellow fever; and such other diseases as are indigenous or epidemical, in the West Indies Islands or in the torrid zone*. In it, he described a syndrome which undoubtedly encompassed malabsorption and which has been claimed by some writers as an early example of tropical sprue; however, careful scrutiny of his book clearly reveals that this was epidemic and hence more likely to have been caused by *Giardia lamblia* infection, or possibly another protozoan infection of the small intestine.[25]

James Grainger (1721–66),[26] who was well known to Samuel Johnson (1709–84), left England for a four-year tour of the West Indies in April 1759. He settled and worked in St Kitts. In 1864 he published an *Essay on the more common West India diseases; and the remedies which that country itself produces; to which are added some hints on the management of negroes*, although it has to be said that his major interest again lay in medical botany.

Little is known about the early life of Colin Chisholm FRS (*c.* 1747–1825);[27] he was a graduate (like many of the early pioneers) of Aberdeen University, and later became a surgeon. It is known that he practised medicine in Grenada and also wrote *An essay on the malignant pestilential fever introduced into the West India islands from Boullam on the coast of Guinea, as it appeared in 1793 and 1794*, first published in 1795. He later wrote *A letter to John Haygarth, MD, exhibiting further evidence of the infectious nature of the pestilential fever in Grenada … and in America*, in 1809.

William Wright FRS (1735–1819)[28] was a military physician who sailed for the West Indies (under Rodney) in 1760 and served in the eastern Caribbean – most notably Martinique. After five-and-a-half years in the Royal Navy, he was paid off in September 1763. Having obtained additional medical qualifications in Britain, he embarked for Jamaica in early 1764, and wrote on yaws. Wright became Surgeon-General of Jamaica in 1774 and returned to Britain three years later, having by then served there for thirteen years. He was also essentially a medical botanist, and discovered *Cinchona jamaicensis*, the basis of many modern anti-malarial drugs (see below). In the summer of 1780, Wright sailed again for Jamaica to become regimental surgeon to the Jamaica regiment; this took two attempts, as his first ship was captured. In 1796 he again returned to the Caribbean, his fourth visit, and was both Physician to the Army and Director of Military Hospitals in Barbados from April 1796 until April 1798. Together with several others, he revised Grainger's (see above) *An Essay on the more common West India diseases…*, which he considered a model work on diseases of black

FIGURE 1.9   James McGrigor (1771–1858) made early contributions to military medicine in the Caribbean (reproduced courtesy of The Wellcome Library, London).

slave populations. He also wrote a report on diseases common among British troops in the West Indies.

Sir James McGrigor (1771–1858; Figure 1.9)[29] became an Army Surgeon (later achieving great eminence) early in his career. In 1795 his regiment sailed for the West Indies and, after a series of disasters, he ultimately arrived and served in Grenada and St Vincent. He wrote *A Report of Sickness, Mortality and Invaliding in the Army in the West Indies* (1838).

India

A book written by William Twining (1790–1835) (see above) in 1834 broke with tradition; this brought him into line with Roupell's classification of diseases while serving with the Seamen's Hospital Society (SHS). Instead of listing entities by symptoms, his knowledge of anatomy led to categorization of diseases according to their effects on various organs, such as the spleen, liver, etc.; his nosology of disease was thus similar to that of Roupell. This book therefore straddled two eras in tropical hygiene and medicine. He 'accepted some of the precepts and assumptions of those who emphasized *climate* [my italics] as a determining factor … [his writing also reflected] the growing impact of a rationalized pathology … in which racial differences [explained] susceptibility or resistance to disease, as well as the appropriateness of particular therapies' – for example,

FIGURE 1.10 James Ranald Martin (1793–1874) made early contributions regarding diseases in India (reproduced courtesy of The Wellcome Library, London).

that aggressive therapies (mercury and bleeding) should not be used on Indians, because their constitutions were less resilient than those of Europeans. Twining used ipecacuanha (and not salivation) for dysentery, and was certainly not in favour of mercury!![30]

James Johnson (1777–1845)[31] published *The Influence of Tropical Climates on European Constitutions* in 1812 (sixth edition, 1841). This encapsulated the growing pessimism among Europeans about their ability to colonize India and adapt to its climate. The book was also 'receptive to Indian culture, some aspects of which – such as diet and dress – were recommended to Europeans inhabiting hot climates'. Johnson challenged prevailing opinions regarding the nature of cholera and other diseases; he also advised on the outbreak which swept Britain in 1831–2.

In 1819 Sir James Ranald Martin, FRCS (1793–1874; Figure 1.10)[32] took charge of the Ramgarh battalion during operations against the hill-tribes in northern Bengal, and this experience 'led him to propose measures to prevent disease in the battalion, including the siting of camps in more elevated and "healthy" positions'. The campaign also impressed on him 'the importance of the *environment* [my italics] in the causation of disease'. He had a lifelong interest in 'medical topography' (see *Medical topography of Calcutta*, 1837), public health and 'sanitary matters'. He also edited the seventh edition of Johnson's book in 1856, and gave a series of lectures on diseases of tropical climates (see *The Lancet* 1850; ii: 145–7, 291–4). In 1857 (i.e. after the Crimean War) Martin was appointed to the Royal Commission to inquire into sanitary conditions in the British Army and, in 1859, the Indian Army.

Sir George Ballingall's (1780–1855)[33] biographer wrote: 'Conditions in India confirmed his belief that most diseases and illnesses could be avoided or at least ameliorated through preventive [my italics] measures'. It was the lifestyles of the Europeans, rather than local conditions in India, which were at fault. He wrote *Practical observations on Fever, Dysentery and Liver Complaints, as they occur amongst the European troops in India* (1818), and *Essay on Syphilis* (1830). Later, in Scotland, Ballingall wrote several books on military surgery. He probably made no significant contributions to medical theory, his major contribution being to military hygiene. 'His belief in the ability of Europeans to acclimatize to Indian conditions was becoming less popular as the [nineteenth] century wore on.' Ballingall was also against the use of mercury.

Charles Morehead FRCP (1807–82)[34] was a supporter of medical education for Indians (of which he was a pioneer) and of the establishment of the board of native education; in 1840 he was appointed its secretary, a post he held for five years. Morehead was the originator of the Grant Medical College, Bombay, built in 1845 as a memorial to Grant. He was appointed First Principal, and First Professor of the Principles and Practice of Medicine, and was a physician to the Sir Jamsetjee Jeejeebhoy Hospital. He founded the Medical & Physical Society of Bombay, of which he later became President (1837–59). Morehead wrote *Researches on the Diseases of India* (1856) before returning to Europe in 1859.

Edmund Parkes FRS (1819–76)[35] was essentially a hygienist. He spent three years on tour with the army in Burma and India, where he experienced cholera, hepatitis and dysentery. His books included *The Dysentery and Hepatitis of India* (1846), and *Researches into the Pathology and Treatment of Asiatic or Algide Cholera* (1847). Parkes later investigated the contagiousness of cholera for the General Board of Health (GBH), in the second London cholera outbreak of 1849; the GBH had felt that 'cholera was primarily caused by filth and poor sanitation', there being no person-to-person transmission. He returned to military medicine following the Crimean War, and was based at Fort Pitt.

Sir Joseph Fayrer FRCP, FRCS, FRS (1824–1907; Figure 1.11)[36] held several surgical appointments in India, and left in 1872. He returned in 1875, accompanying the Prince of Wales (later King Edward VII, 1841–1910). His research was largely on snake venoms (1867); he wrote *The Thanatophidia of India* (1872). Fayrer was essentially an epidemiologist whose interests centred on climate, Indian diseases and sanitation.

James Annesley (1780–1847; Figure 1.12) is best known for his two-volume, beautifully illustrated quarto work on the local diseases of Bengal.

Joseph Fayrer

FIGURE 1.11   Sir Joseph Fayrer (1824–1907), doyen of the Indian Medical Service (reproduced courtesy of The Wellcome Library, London).

In 1868, Timothy Richard Lewis (1841–86; Figure 1.13)[37] was selected to 'study … current theories of cholera causation'. Lewis, together with Douglas Cunningham, travelled to India in January 1869. In India, his research centred on cholera, leprosy, 'oriental sore', enteric fever, relapsing fever and prison dietaries. As a microscopist, Lewis was pre-eminent; he visualized the microfilaria of *W. bancrofti* in urine, which he termed *Filaria sanguinis hominis*, and, later, the adult male and female forms of the helminth in blood (see Chapter 3). He also demonstrated amoebae in the human intestine.

As far as *preventive* medicine in the tropics is concerned, an early exponent was Henry Vandyke Carter (1831–97).[38] Born in Hull and educated at St George's Hospital, London, he joined the Bombay establishment, becoming Professor of Anatomy and Physiology at the Grant Medical College. His researches were into leprosy (see Chapter 18) and 'spirillum fever'; his initial view was that the former was essentially an inherited disease (see Chapter 18), but, following a visit to Norway to ascertain whether the disease there was identical with that in India, he became convinced that it was a contagious disease. His belief in leper asylums

FIGURE 1.12   James Annesley (1780–1847) provided documentation of disease in Bengal (reproduced courtesy of The Wellcome Library, London).

thereby escalated. Carter later became Deputy Surgeon-General, IMS, and an Honorary Surgeon to Queen Victoria.

Africa

David Livingstone (1813–73) (see above) stands out for his major literary contributions to 'medicine in the tropics' in Africa prior to the 'germ theory'.

SPECIFIC PROPHYLAXIS AND TREATMENT

The pioneers of early prophylaxis and treatment were, in a sense, far ahead of their time; most carried out their researches not in the tropics, but in Britain. We shall never know who introduced the 'bark' for the 'intermittent fevers'. However, the introduction of variolation into Britain is attributed to Mary Wortley Montagu (1689–1762; Figure 1.14).[39] Edward Jenner (1749–1823; Figure 1.15)[40] and the Director of the St Pancras Smallpox Hospital, William Woodville (1752–1805; Figure 1.16),[41] introduced widespread vaccination for variola (smallpox) in the late eighteenth and very early nineteenth centuries – long before the 'germ theory'

FIGURE 1.13 Timothy Lewis (1841–86) carried out important medical research in India (reproduced courtesy of The Wellcome Library, London).

of disease had been widely propagated. Figure 1.17 shows the London Smallpox Hospital following removal from Battle Bridge (King's Cross).

Introduction of the 'bark', and differentiation of the 'intermittent fevers'

The history of the Peruvian bark, which was first known in Europe in about 1630, is of interest; the discoverer remains unknown. A legendary story that Lady Chinchón, wife of the Viceroy of Peru, was cured of a fever and was later supposed to have brought the bark to Spain has been disproved. It seems more likely that the indigenous remedy was discovered by Spanish missionaries who had observed the work of Andean herbalists. Cardinal Juan de Lugo promoted its use in seventeenth-century Spain; following that, the powdered bark (Jesuit's powder), also known as *Pulvic cardinalis* and *Pulvis partum*, became widely used in Europe.[42] First documentation of its use in England was by John Metford of Northampton in 1656, but it was not until 1672 that this new remedy acquired

FIGURE 1.14 Lady Mary Wortley Montagu (1689–1762) probably introduced variolation into Britain (reproduced courtesy of The Wellcome Library, London).

FIGURE 1.15 Edward Jenner (1749–1823) introduced widespread vaccination against smallpox (reproduced courtesy of The Wellcome Library, London).

FIGURE 1.16 William Woodville (1752–1805) performed clinical trials of vaccination at St Pancras (reproduced courtesy of The Wellcome Library, London).

FIGURE 1.17 The London Smallpox Hospital following removal from Battle Bridge (King's Cross) (reproduced courtesy of The Whittington Hospital, London).

FIGURE 1.18   Thomas Sydenham (1624–89) observed the use of the 'bark' in curing the 'intermittent fever' (reproduced courtesy of The Wellcome Library, London).

wide acceptance – the year when Robert Tabor (or Talbor) (1642–81), an apprentice apothecary, successfully treated King Charles II (1630–85) for a persistent ague with his secret concoction of Peruvian bark. Tabor was both knighted and created Royal Physician, and also protected by the King from interference with his London practice. The bark was included in the *London Pharmacopeia* third edition in 1677, as *Cortex peruanus*.[43] Early in the following century (1712), Francesco Torti of Medena was probably the first to indicate that the bark only cured the 'intermittent fever', although Thomas Sydenham (1624–89; Figure 1.18), the 'English Hippocrates', is usually credited with this observation.

In 1742, Carl Linnaeus (1707–78; Figure 1.19),[44] wishing to immortalize the name of Lady Cinchón, mistakenly gave the tree the generic name *cinchona* (a misspelling of her name). He added the name quinquina, which early French

LINNÆUS

FIGURE 1.19   Carl Linnaeus (1707–78) first used the term chinchona for the 'bark' (reproduced courtesy of The Wellcome Library, London).

naturalists gave to the Peruvian balsam plant (*Myroxylon peruiferum*), known since the sixteenth century.

Widespread demand for the new remedy initiated botanical expeditions to distant parts of the New World; these increased when two French chemists isolated the alkaloids quinine and cinchonine from samples of chinchona bark. Efforts were also made to start plantations in other parts of the world – especially India, Ceylon and the Dutch East Indies. Involvement of the Royal Botanical Gardens, Kew, was important, but credit for locating the variety which produced the highest yield of quinine must go to Charles Ledger (1818–1905; Figure 1.20), an English trader in Peru. The bark of *Cinchona ledgeriana* was found to possess a quinine content approaching five times that of the more usual yield. His seeds (which his

FIGURE 1.20   Charles Ledger (1818–1905) initiated the mass commercial production of quinine in south-east Asia (reproduced courtesy of The Wellcome Library, London).

servant collected in an Andean region of Bolivia) were later planted in Java. By the late nineteenth century quinine manufacture had passed into the hands of large profit-making enterprises, and in 1918 the Second Quinine Convention gave the Dutch monopoly of the quinine industry. After the Japanese occupation of Indonesia during the Second World War (1939–45) the emphasis fell on synthetic antimalarials, manufacture of which had begun in the 1930s with the German discovery of mepacrine (see Chapter 15). With increasing drug resistance in *Plasmodium falciparum*, the fever bark has recently seen a recrudescence.[46]

THE BIRTH OF 'FEVER' HOSPITALS

Fever hospitals grew up in Britain in the early nineteenth century – Haygath's in Chester being an early example. The use of chinchona, however, never became as firmly established in the British-based medical establishment as it did among colleagues with clinical experience in the tropics. Its inflammatory element in fever was also much debated.[47]

FIGURE 1.21   John Snow (1813–58) hypothesized that cholera was transmitted by human faeces (reproduced courtesy of The Wellcome Library, London).

EARLY ATTEMPTS AT UNDERSTANDING DISEASE TRANSMISSION

Recognition that a 'contaminated environment' is conducive to disease propagation was suggested by many tropical physicians, and pre-dated the 'germ theory'. John Snow (1813–58; Figure 1.21)[48] and William Budd (1811–90; Figure 1.22)[49] suggested that certain gastrointestinal diseases were transmitted by the faecal route; before them, Chadwick and Southwood Smith[50] had, amongst others, highlighted the 'filth' concept of diseases. In the navy, it was well known before the advent of 'germ theory' that hygienic factors were involved in the transmission of typhus; Thomas Spencer-Wells (1818–97)[51] contributed greatly to naval hygiene.

FIGURE 1.22   William Budd (1811– 90) provided evidence that typhoid is transmitted by human faeces (reproduced courtesy of The Wellcome Library, London).

NOTES

1 *Miasma*: 'Infectious or noxious exhalations from putrescent organic matter; poisonous particles or germs floating in and polluting the atmosphere; noxious emanations, esp. malaria'. J A Simpson, ESC Wiener (eds). *The Oxford English Dictionary*, 2nd edn, Vol. 9. Oxford, 1989: Clarendon Press, p. 710.

2 *Contagious*: 'Communicable or infections by contact'. J A Simpson, ESC Wiener (eds). *The Oxford English Dictionary*, 2nd edn, Vol. 3. Oxford, 1989: Clarendon Press, p. 806.

3 C Lloyd, J L S Coulter. The West African Squadron. In: *Medicine and the Navy 1200–1900*, Vol. 4. London, 1963: E & S Livingstone, pp. 155–72; L Rogers. *Fevers in the Tropics*. Oxford, 1908: Oxford University Press. See also H Thomas. *The Slave Trade: the history of the Atlantic slave trade 1440–1870*. London, 1907: Picador, p. 925.

4 C Lloyd, J L S Coulter. Fevers. In: *Medicine and the Navy 1200–1900*, Vol. 4. London, 1963: E & S Livingstone, pp. 173–96.

5 G C Cook. Doctor David Livingstone FRS (1813–1873): 'the fever' and other medical problems of mid-nineteenth century Africa. *J Med Biog* 1994, 2: 33–43; A D Roberts. Livingstone, David (1813–1873). In: H C G Matthew, B Harrison (eds) *Oxford Dictionary of National Biography*, Vol. 34. Oxford, 2004: Oxford University Press, pp. 73–82.

6 John Atkins (*c.* 1685–1757) was a naval surgeon. As Surgeon of the *Tartar* (1705–10) he saw action against the French in the Channel, and he served on the bomb-ketch *Lion* in the Mediterranean from 1710 to 1714. Later he served on the *Swallow*, a ship dispatched to seize pirates operating on the West African coast; here, he saw much tropical diseases and described *trypanosoma gambiense* infection, ophthalmic onchocerciasis, dracontiasis and cerebral malaria. He later wrote *The Navy Surgeon, or Practical System of Surgery* (1732) and *A Voyage to Guinea, Brasil and the West Indies* (1735). He also denounced the slave trade, which he had witnessed at first hand on the West African coast, and he might have served for a short time on a slave-ship. See also J Watt. Atkins, John (bap.1685, d.1757). In: H C G Matthew, B Harrison (eds), *Oxford Dictionary of National Biography*, Vol. 2. Oxford, 2004: Oxford University Press, pp. 819–20.

7 George Playfair (1782–1846) qualified MRCS in 1802. He was a Surgeon's Mate on *Ocean* (1801–2), and later Surgeon on the same ship (1802–4). He became Assistant Surgeon (21 March 1805), and Surgeon (22 February 1817). He served on steam ships from 20 February 1832, and became a member of the Medical Board and Chief Inspector-General of Hospitals in Bengal (31 December 1842). He retired on 1 March 1843. His four sons were Lyon (later first Baron Playfair) (1818–98), Robert (later Lt Col Sir) (1818–99), George Rankin Playfair (1816–81) and William Smoult Playfair (1835–1903). He served in Afghanistan (1839–41) and was the author of *Taleef Shareef* and the Indian *Materia Medica* (1833). See also D G Crawford. Playfair, George. *Roll of the Indian Medical Service 1615–1930*. Calcutta, 1930: W Thacker & Co, p. 52; G J N Gooday. Playfair, Lyon, First Baron Playfair (1818–1898). In: H C G Matthew, B Harrison (eds), *Oxford Dictionary of National Biography*, Vol. 44. Oxford, 2004: Oxford University Press, pp. 556–60; E J Carlyle, L Milne. Playfair, Sir Robert Lambert (1828–1899). In: H C G Matthew, B Harrison (eds), *Oxford Dictionary of National Biography*, Vol. 44. Oxford, 2004: Oxford University Press, pp. 561–2; A Dally. Playfair, William Smoult (1835–1903). In: H C G Matthew, B Harrison (eds), *Oxford Dictionary of National Biography*, Vol. 44. Oxford, 2004: Oxford University Press, pp. 566–7; Anonymous. Playfair, William Smoult (1835–1903). *Munk's Roll*, Vol. 4. London: Royal College of Physicians, pp. 184–5.

8 Anonymous. Early works on tropical medicine. *Br Med J* 1912, ii: 1481; Anonymous. One hundred years ago: the knowledge of tropical diseases in 1813. *Br Med J* 1913, i: 455–6; Anonymous. Early references to tropical medicine. *Br Med J* 1913, i: 1331–2; C Singer. Notes on some early references to tropical diseases. *Ann Trop Med Parasitol* 1912, 6: 87–101, 379–402. See also C Playfair. On the good effect of a combination of ipecacuan and landanum in dysentery. *Edin Med Surg J* 1813, 9: 18–22; F Tymon. Cases of hydrophobia, treated, one of them successfully. *Edin Med Surg J* 1813, 9: 22–30; J Shoolbred. Case of hydrophobia, successfully treated. *Edin Med Surg J* 1913, 9: 30–49; A Naval Surgeon. On the introduction of the depletory method of cure in tropical fever. *Edin Med Surg J* 1913, 9: 51–3; J H Dickson. Extracts from a circular letter to the surgeons of his Majesty's Ships and Vessels on the Leeward Islands Station. *Edin Med Surg J* 1913, 9: 53–8.

9 G C Cook. Scott, Sir (Henry) Harold (1874–1956). In: H C G Matthews, B Harrison (eds), *Oxford Dictionary of National Biography*, Vol. 49. Oxford, 2004: Oxford University Press, pp. 386–7; G C Cook. Joseph William Bazalgette (1819–1891): a major figure in the health improvements of Victorian London. *J Med Biog* 1999, 7: 17–24; G C Cook. Construction of London's Victorian sewers: the vital role of Joseph Bazalgette. *Postgrad Med J* 2001, 77: 802–4.

10 G C Cook, A J Webb. Perceptions of malaria transmission before Ross' discovery in 1897. *Postgrad Med J* 2000, 76: 738–40.

11 H H Scott. *A History of Tropical Medicine*, two volumes. London, 1939: Edward Arnold, pp. 648, 1165.

12 James Lind (1718–94) was the pioneer of nautical hygiene, who is credited with the first clinical trial of citrus fruits in scurvy. See also M Bartholomew. Lind, James (1716–1794). In: H C G Matthew, B Harrison (eds), *Oxford Dictionary of National Biography*, Vol. 33. Oxford, 2004: Oxford University Press, pp. 810–13.

13 Sir John Pringle (1707–82) was a pioneer of military hygiene. See also J S G Blair. Pringle, Sir John, first baronet (1707–1782). In: H C G Matthew, B Harrison (eds), *Oxford Dictionary of National Biography*, Vol. 45. Oxford, 2004: Oxford University Press, pp. 398–400.

14 William Cullen (1710–90) was an eminent Scottish physician. See also W F Bynum. Cullen, William (1710–1790). In: H C G Matthew, B Harrison (eds), *Oxford Dictionary of National Biography*, Vol. 14. Oxford, 2004: Oxford University Press, pp. 581–6; W F Bynum. Cullen and the study of fevers in Britain, 1760–1820. In: W F Bynum, V Nutton (eds), *Theories of Fever from Antiquity to the Enlightenment*. London, 1981: Wellcome Institute for the History of Medicine, pp. 135–47.

15 John Coakley Lettsom (1744–1815) was an eminent physician and philanthropist who founded the Medical Society of London in 1773. See also J F Payne, R Porter. Lettsom, John Coakley (1744–1815). In: H C G Matthew, B Harrison (eds), *Oxford Dictionary of National Biography*, Vol. 33. Oxford, 2004: Oxford University Press, pp. 513–15.

16 G C Cook. Robert Robertson, FRS (1742–1829): physician to the Royal Hospital, Greenwich, 18th century authority on 'fever', and early practitioner in care of the elderly. *J Med Biog* 2006, 14: 42–45.

17 G C Cook. George Leith Roupell FRS (1797–1854): significant contributions to the early nineteenth century understanding of cholera and typhus. *J Med Biog* 2000, 8: 1–7. See also G C Cook. Medical disease in the Merchant Navies of the World in the days of sail: the Seamen's Hospital Society's experience. *Mariner's Mirror* 2005, 91: 46–51.

18 F Riollay. *Critical Introduction to the Study of Fevers*. London, 1788: T Cadell.

19 *Op cit.* See note 14 above.

20 J Boyle. *A Practical Medico-historical Account of the Western Coast of Africa*. London, 1831: Highley.

21 J Jones. *Medical and Surgical Memoirs, 1855–1890*. New Orleans, (1876–90): J Jones.

22 C M Poser, G W Bruyn. *An Illustrated History of Malaria*. London, 1999: Parthenon Publishing Group, p. 165. See also P Schlagenhauf. Malaria: from prehistory to present. *Infect Dis Clin North Am* 2004, 18: 189–205.

23 G C Cook. William Twining (1790–1835): the first accurate clinical descriptions of 'tropical sprue' and kala-azar? *J Med Biog* 2001, 9: 125–31; D M Peers. Twining, William (1790–1835). In: H C G Matthew, B Harrison (eds), *Oxford Dictionary of National Biography*, Vol. 55. Oxford, 2004: Oxford University Press, pp. 729–30.

24 A MacGregor. Sloane, Sir Hans, baronet (1660–1753). In: H C G Matthew, B Harrison (eds), *Oxford Dictionary of National Biography*, Vol. 50. Oxford, 2004: Oxford University Press, pp. 943–9; A MacGregor (ed.). *Sir Hans Sloane: collector, scientist, antiquary*. London, 1994: British Museum, p. 308.

25 C C Booth. Hillary, William (1697–1763). In: H C G Matthew, B Harrison (eds), *Oxford Dictionary of National Biography*, Vol. 27. Oxford, 2004: Oxford University Press, pp. 212–13; G C Cook. 'Tropical sprue': some early investigators favoured an infective cause, but was a coccidian protozoan involved? *Gut* 1997, 40: 428–9.

26 G Goodwin. C Overy. Grainger, James (1721–1766). In: H C G Matthew, B Harrison (eds), *Oxford Dictionary of National Biography*, Vol. 23. Oxford, 2004: Oxford University Press, pp. 259–60.

27 G Goodwin, J S Reznick. Chisholm, Colin (d.1825). In: H C G Matthew, B Harrison (eds), *Oxford Dictionary of National Biography*, Vol. 11. Oxford, 2004: Oxford University Press, pp. 485–6.

28 N D Saakwa-Mante. Wright, William (1735–1819). In: H C G Matthew, B Harrison (eds), *Oxford Dictionary of National Biography*, Vol. 60. Oxford, 2004: Oxford University Press, pp. 503–5.

29 H M Chichester, J S G Blair. McGrigor, Sir James, first baronet (1771–1858). In: H C G Matthew, B Harrison (eds), *Oxford Dictionary of National Biography*, Vol. 35. Oxford, 2004: Oxford University Press, pp. 447–9.

30 *Op cit.* See note 23 above.

31 A Greenhill, M Harrison. Johnson [Johnstone], James (1777–1845). In: H C G Matthew, B Harrison (eds), *Oxford Dictionary of National Biography*, Vol. 30. Oxford, 2004: Oxford University Press, pp. 274–5.

32 M Harrison. Martin, Sir James Ranald (1793–1874). In: H C G Matthew, B Harrison (eds), *Oxford Dictionary of National Biography*, Vol. 36. Oxford, 2004: Oxford University Press, pp. 945–6.

33 D M Peers. Ballingall, Sir George (1780–1855). In: H C G Matthew, B Harrison (eds), *Oxford Dictionary of National Biography*, Vol. 3. Oxford, 2004: Oxford University Press, pp. 596–7.

34 A J Arbuthnot, C E J Herrick. Morehead, Charles (1807–1882). In: H C G Matthew, B Harrison (eds), *Oxford Dictionary of National Biography*, Vol. 39. Oxford, 2004: Oxford University Press, pp. 81–2.

35 M Harrison. Parkes, Edmund Alexander (1819–1876). In: H C G Matthew, B Harrison (eds), *Oxford Dictionary of National Biography*, Vol. 42. Oxford, 2004: Oxford University Press, pp. 762–4.

36 H P Cholmeley, W F Bynum. Fayrer, Sir Joseph, first baronet (1824–1907). In: H C G Matthew, B Harrison (eds), *Oxford Dictionary of National Biography*, Vol. 19. Oxford, 2004: Oxford University Press, pp. 200–202.

37 M Pelling. Lewis, Timothy Richards (1841–1886). In: H C G Matthew, B Harrison (eds), *Oxford Dictionary of National Biography*, Vol. 33. Oxford, 2004: Oxford University Press, pp. 659–60.

38 M Harrison. Carter, Henry VanDyke (1831–1897). In: H C G Matthew, B Harrison (eds), *Oxford Dictionary of National Biography*, Vol. 10. Oxford, 2004: Oxford University Press, pp. 350–51.

39 I Grundy. Montagu, Lady Mary Wortley (bap. 1689, d. 1762). In: H C G Matthew, B Harrison (eds) *Oxford Dictionary of National Biography*, Vol. 38. Oxford, 2004: Oxford University Press, pp. 754–9.

40 D Baxby. Jenner, Edward (1749–1823). In: H C G Matthew, B Harrison (eds), *Oxford Dictionary of National Biography*, Vol. 30. Oxford, 2004: Oxford University Press, pp. 4–8.

41 D Brunton. Woodville, William (1752–1805). In: H C G Matthew, B Harrison (eds), *Oxford Dictionary of National Biography*, Vol. 60. Oxford, 2004: Oxford University Press, pp. 230–31. See also: G C Cook. Dr William Woodville (1752–1805) and the St Pancras Smallpox Hospital. *J Med Biog* 1996, 4: 71–8.

42 F. Guerra. The introduction of Chinchona in the treatment of malaria. *J Trop Med Hyg* 1977, 80: 112–18, 135–9.

43 J Jeramillo-Arango. *The Conquest of Malaria*. London, 1950: Heinemann.

44 Carl Linnaeus (Carl von Linné) is widely known for introducing the present system of botanical classification. See also B D Jackson. *Linnaeus*. London, 1923: H F & G Witherby, p. 416; H Goerke. *Linnaeus*. New York, 1973: Charles Scribner's Sons, p. 178; *Carl von Linné*. Leipzig, 1978: B G Teubner, p. 124; L Koener. *Linnaeus: Nature and Nation*. Cambridge, 1999: Harvard University Press, p. 298; W Blunt. *Linnaeus: the compleat naturalist*. London, 2004: Frances Lincoln Ltd, p. 287.

45 G Gramiccia. *The Life of Charles Ledger (1818–1905): alpacus and quinine*. London, 1988: Macmillan Press, p. 222.

46 C R Markham. *Peruvian Bark: a popular account of the introduction of chinchona cultivation into British India*. London, 1880: John Murray, p. 550; N Taylor. *Chinchona in Java: the story of quinine*. New York, 1945: Greenberg, p. 87; M L Duran-Reynals. *The Fever Bark Tree: the pageant of quinine*. London, 1947: W H Allen, p. 251; L J Bruce-Chwatt. Three hundred and fifty years of the Peruvian bark. *Br Med J* 1988, i: 1486–7.

47 *Op cit*. See note 14 above (Bynum).

48 A E Shephard. *John Snow: anaesthetist to a queen and epidemiologist to a nation. A biography*. Cornwall, 1995: York, p. 373; S J Snow. Snow, John (1813–1858). In: H C G Matthew, B Harrison (eds), *Oxford Dictionary of National Biography*, Vol. 51. Oxford, 2004: Oxford University Press, pp. 495–8. See also S Johnson. *The Ghost Map*. London, 2006: Allen Lane.

49 M Pelling. Budd, William (1811–1880). In: H C G Matthew, B Harrison (eds), *Oxford Dictionary of National Biography*, Vol. 8. Oxford, 2004: Oxford University Press, pp. 551–2; M Dunnill, Dr William Budd: Bristol's most famous physician. Bristol 2006: Redcliffe Press, p. 158.

50 G C Cook. Thomas Southwood Smith FRCP (1788–1861): leading exponent of diseases of poverty and pioneer of sanitary reform in the mid-nineteenth century. *J Med Biog* 2002, 10: 194–205.

51 G C Cook. Thomas Spencer Wells, Bt, FRCS (1818–97) and his contributions to Naval Medicine. *J Med Biog* 2007, 15: 63–7.

2

Origins of the formal discipline: background factors

In the introduction to his textbook *Tropical Diseases* (the twenty-second edition of which is presently in preparation), published in 1898, Manson wrote:

> The title I have elected to give to this work, TROPICAL DISEASES, is more convenient than accurate. If by 'tropical diseases' be meant diseases peculiar to, and confined to, the tropics, then half a dozen pages might have sufficed for their description ... . If ... the expression 'tropical diseases' be held to include all diseases occurring in the tropics, then the work would require to cover almost the entire range of medicine.

And in his inaugural lecture to the London School of Tropical Medicine (LSTM) on 2 October 1899, he explained:

> In the main, the etiology of disease is but a branch of natural history. Climate, that is, temperature, influences pathology mainly, if not only, inasmuch and so far as it influences the distribution of the pathogenic flora and fauna which, just as in the case of the ordinary fauna and flora, are markedly regulated by atmospheric conditions ... . The geographical limitations of the animal parasitic diseases are undoubtedly, in many instances, determined by atmospheric temperature. But although high temperature may be an indispensable and ultimate determining factor in their distribution, temperature does not usually operate directly on the causal germ; its operation is usually an indirect one, acting probably through many channels ... .

In early 1898 Joseph Chamberlain (1836–1914; see Chapter 20), probably at Manson's suggestion, wrote to the Committee of Management of the Seamen's

Hospital Society requesting the establishment of a School of Tropical Medicine at the Branch (Albert Dock) Hospital (see Chapter 3). This British event undoubtedly gave the initial impetus for the formation of tropical medicine as a distinct discipline. However, a new enterprise can only come into being if the *milieu* is well prepared for its reception. What, therefore, were the underlying factors that formed the backdrop for this new discipline?

In recent years it has become customary amongst historians of medicine to equate '*tropical* medicine' with '*colonial* medicine'. There is certainly some justification for this; the discipline blossomed during that period when the Empire was at its zenith (see Chapter 1), and a knowledge of the diseases which afflicted the servants of the Raj and Empire was, in the eyes of the politicians, of paramount importance. Tropical medicine was in fact an integral component in Joseph Chamberlain's plan for 'constructive imperialism'.[1]

It is doubtless more accurate, therefore, to envisage colonial politics as exploiting the newly established discipline for its own ends. The concept of an Empire-dominated discipline also ignores several background factors which lay behind the development of tropical medicine in the latter years of the nineteenth and the beginning of the twentieth century. The scenario into which tropical medicine and, shortly afterwards, the (Royal) Society of Tropical Medicine and Hygiene were born included:

- The public health (or hygiene) movement – both in Britain and later India
- Travel – the era of African exploration, latterly replaced by travel, was a dominant theme in the contemporary media
- Natural history
- Evolutionary theory, and the controversy surrounding 'Darwinism'
- Escalating knowledge of disease causation – the 'germ theory'
- Rapid advance(s) in clinical parasitology.

THE PUBLIC HEALTH MOVEMENT

Britain

In Britain, an improvement in sanitation and hygienic standards had begun soon after the mid-eighteenth century; initially this movement was most advanced in the Army and the Royal Navy (RN). In the cities, it coincided with increased population density, which began with the Industrial Revolution (people were moving *en masse* from the country). Also, the population of Britain increased significantly after 1740 and, despite contrary views, this was associated with a general improvement in living standards and a reduction in mortality (especially in childhood); some (infective) diseases, such as malaria, decreased in prevalence. However, water supplies and sanitation remained seriously flawed – for example, by 1830 water-closets were not widely installed, even in the better houses, while in London cesspools were still in use at the middle of the century.

By the mid-eighteenth century, the 'hospital and dispensary movement' had come into existence; several hospitals, including St Bartholomew's and the London, were rebuilt. Following this, the general state of hospitals remained essentially unchanged until the second half of the nineteenth century, when significant advances in nursing and surgical care came about. Certain diseases, such as plague and smallpox, had become better understood in the latter days of the eighteenth century; the initiation of quarantine regulations and the wide-spread vaccination introduced by Edward Jenner (1749–1823; see Chapter 1) had contributed significantly.[2]

The 'nation's conscience' was awakened in the early nineteenth century by Jeremy Bentham (1748–1832); his underlying philosophy was that all factors influencing the health of the country should be the concern of the legislature. Bentham's Utilitarian Philosophy attempted in fact to draw a parallel between the physical and social sciences. This arousal of public awareness of social problems was highlighted during Britain's first cholera epidemic (during which Bentham died) in 1831–2. Cholera has arguably been termed the 'great sanitary reformer', although it is probably more accurate to substitute 'fever' for cholera (see Chapter 1); John Snow (1813–58), the London anaesthetist, established a link between faecally-contaminated drinking water and this disease in two important publications in 1849 and 1854. This should be taken in conjunction with Charles Dickens' (1812–70) works (especially *Oliver Twist*, 1837) and the Reverend Charles Kingsley's (1819–75) best-known book *The Water Babies* (1863); in fact, the latter's views were so outspoken that he was banned by the Bishop of London from preaching in any of the City's pulpits!

'Public health' was therefore 'an affair of legislation and administration'. Edwin Chadwick (1800–90), a barrister and journalist and a close associate of Bentham, developed 'the sanitary idea', while most medical input came from Neil Arnott (1788–1874), Thomas Southwood-Smith (1788–1861), and James Kay (later Kay-Shuttleworth) (1804–77). Chadwick was appointed Secretary to the Poor Law Commission, and maintained that disease was closely influenced by 'external sanitation and drainage'; his revolutionary work was *Report on the Sanitary Conditions of the Labouring Population* (1842). Chadwick initiated an investigation into the 'fever' (most cases were caused by epidemic typhus), the work being carried out by Arnott and Kay-Shuttleworth, both of whom were 'miasmatists'. In the mid-nineteenth century, infectious disease was considered to be caused by filth, and therefore the 'sanitary revolution' consisted of a battle against dirt and nuisance(s). The reasoning behind this was clearly flawed; the 'germ theory' (see below) had yet to be formulated.

A Royal Commission, the 'Health of Towns Commission', the main objective of which was to apply this report, was created. Medical Officers of Health were appointed in Liverpool (1842) and London (1848); Sir John Simon (1816–1904), in the latter post, subsequently produced eight outstanding reports on living and health conditions in the metropolis. Another of Simon's 'advances' was to scrap the eighteenth-century quarantine regulations and to institute (via shipping) preventive strategies for both 'native' and 'foreign' infections. A General Board of Health was set-up in 1848; this included: Southwood-Smith (see above) and Ashley-Cooper (later the seventh Earl of Shaftesbury) (1801–85). Chadwick's career was thus dominated by the foundation of the public health movement in Britain; however, as Wilcocks has pointed out, his work originated not from a medical but from an economic standpoint. Chadwick considered that 'disease – especially typhus – caused a drain on public expenditure for the relief of individual patients; so high was the cost that it seemed essential to remove the *cause* of the disease, and thus relieve public expenditure'.[3]

The two major environmental hazards of the mid-nineteenth century originated in atmospheric and river pollution. The former gave rise to the 'Great Stink' of 1858 (which aroused a great deal of debate in, and temporary closure of, the Houses of Parliament). The River Thames, the source of drinking water for the metropolis, had become heavily contaminated with sewage – a matter which gave rise to a campaign by the satirical magazine *Punch* to bring the matter to public attention. The Thames had also become heavily contaminated with all manner of other debris – including dog, cat and pig carcasses. With the introduction of (Sir) Joseph Bazalgette's (1819–91) sewers and the building of the Thames Embankment, the situation began to improve. In 1875 the Public Health Act (the public health charter for the next 60 years) was introduced, and the Metropolitan Water Board was established in 1902.

Not until the early days of the twentieth century, however, was the health of the individual to become a dominant theme, heralded by the 1911 National Health Insurance Act. Areas of importance were school medical services, maternal and child welfare, tuberculosis control, monitoring of infectious disease(s), and the quality of domestic water supplies.[4]

India

A rapid expansion in the expatriate population of India was occurring in the early nineteenth century; here, disease was rife, and British subjects were certainly not exempt from local (infectious) diseases. Following upon developments in public health in Britain, strenuous efforts were made to institute similar ones in India. A Sanitary Act was passed in 1864, and a Town Improvement Act the following year. These developments undoubtedly formed an important backdrop to the development of the *tropical medicine* specialty. The origins of the Sanitary Organizations of India have been summarized: the Royal Sanitary Commission considered that if measures were applied to India under an organized Public Health Service, health there would soon reach an acceptable standard; in 1864, a Sanitary Commissioner was appointed for each Province, with special health officers for Calcutta, Bombay and Madras. William Simpson (1855–1931; see Chapter 14) concluded in 1919 that:

> The root of the matter is there is no Minister of Health in India, whose duty is concerned solely with sanitation and the health of the population [and] until a proper sanitary service for India

is formed – which would be a splendid outlet for young Indian medical men and women – there will be no real progress, and India will remain defenceless against epidemics.

Following her nursing work at Scutari, Florence Nightingale (1820–1910) also addressed the public health problems of India. Although she did not actually visit the country, she wrote *Observations on the Sanitary Conditions of the British Army in India* (1863).[5]

Other contributions to the public health debate

Simpson also addressed the progress of sanitation in West Africa. Here, he advocated a

> great Colonial Medical and Sanitary Service, the members of which, well trained in tropical diseases and their prevention in our Tropical Schools, should be available after good service and experience in one colony for transfer to others if required. There should, he continued, be a local service of native medical men and sanitary inspectors who have been trained locally, and whose knowledge … would be of great assistance to the hospital physician, sanitary officer, and Government in their health administration and their campaign against disease.

Sir Ronald Ross (1857–1932; see Chapter 5) addressed hygienic/sanitary conditions in the tropics in 1909. Hygiene and sanitation in tropical countries were also dominant themes in several lectures delivered in the early years of the London School of Tropical Medicine (LSTM) and elsewhere; much of the address by Sir William MacGregor (1847–1919) given to the LSTM on 3 October 1900 (the first anniversary of that institution) dwelt on these issues, as did that given at the tenth anniversary by Sir William Osler (1849–1919) on 26 October 1909. Sir Andrew Balfour (1873–1931), first (and last) Director of the London School of Hygiene and Tropical Medicine (LSHTM), also dealt with problems underlying hygiene and sanitation in the tropics, while Manson (in 1898) had been only too well aware that much of the disease commonly encountered in tropical countries resulted from a highly contaminated environment:[6]

> Certain cosmopolitan diseases, such as leprosy and plague … have been practically ousted from Europe and the temperate parts of America by the spread of civilization, and the improved hygiene that has followed in its train, and are now practically confined to tropical and sub-tropical countries, where they still survive under those backward social and unsanitary conditions which are necessary for their successful propagation ….

TRAVEL – THE VICTORIAN EXPLORERS

Travel was a dominant theme in Victorian Britain and gained an extensive coverage in the media, as does space travel today. David Livingstone (1813–73; see Chapter 1) stood out above all other great explorers – 'Commerce, Christianity and Civilisation for Africa' were his goals. Others included in a long and distinguished list of African travellers were: James Bruce, Mungo Park, Clapperton, the Landers, Burton, Speke, Grant and Baker.

There were also lesser-known explorers, and these included several who were to become eminent scientists of their day – Charles Darwin, Alfred Wallace and Thomas Huxley. The great voyages of exploration undertaken by Darwin (1809–82) and Huxley (1825–95) on *HMS Beagle* and *HMS Rattlesnake*, respectively, had a profound influence on their subsequent thoughts. In southern Africa, Francis Galton (1822–1911) – a cousin of Darwin, and best known for his magnum opus *Hereditary Genius* (1869) – was also a major figure. In southern America, the adventures of Charles Waterton (1782–1865) also caught the public's imagination in *Wanderings in South America* (1826).[7]

NATURAL HISTORY

By the mid-nineteenth century, natural history had become virtually a national obsession. Books on the subject were only marginally less popular than the novels of Dickens, and an undistinguished natural history book by the Reverend J G Wood, *Common Objects of the Country*, sold 100 000 copies weekly! At the beginning of the century, the subject had not only been neglected and under taught in schools until the 1880s, but was also positively despised; Kingsley recalled that the naturalist was regarded as 'a harmless enthusiast, who went "bug-hunting" simply because he had not the spirit to follow a fox'!

Gilbert White's (1720–93) *Natural History of Selborne* (1789) remained an important source of inspiration for every amateur Victorian naturalist. Sir Joseph Banks (1744–1820) was accepted as a great eighteenth-century botanist/explorer. Looking back further, the Swede Carl Linnaeus (1707–1778; see Chapter 1) had launched the 'modern era' of natural history by introducing his system of classification. It is likely that the popularity of natural history was associated with a lack of serious scientific advance in the subject; it is easier, for example, for a layman to comprehend a subject which is not undergoing a revolution(s)!

The prime purpose of the study of nature was to assume a closer relationship with God; the Reverend William Paley (1743–1805) wrote a detailed exposition in his *Natural Theology* (1802). The main *raison d'être* was to teach of God's existence, whilst the second was to illustrate His attributes. Discovery of a new species was the highest goal a naturalist could achieve, while the construction of theories was shunned. Study of natural history became, therefore, virtually a pious duty, and until the arrival of Darwin, Huxley and Tyndell, science did not have a champion who was prepared to compete with religion. This was a period of the great illustrators, of whom John Anderson and John Gould were to the fore.[8]

Visits to the new zoological and botanical gardens and public aquaria, forays into rock pools, magic lantern lectures and evenings at the microscope all filled in the interminable Victorian leisure hours. The Zoological Society of London was founded by Sir Stamford Raffles (1781–1826) (founder of Singapore) in 1826; he also became its first Director. Two years later, London's Zoological Gardens opened. Seaweed albums, butterfly cases and stuffed birds became dominant features of the average Victorian drawing room. Many individuals collected shells, seaweeds, ferns, fossils and butterflies, amongst many other 'natural' things. Botanizing and botany were considered to be women's hobbies, whilst zoology remained a male-dominated interest. The introduction of binoculars and photography allowed animal behaviour to be objectively recorded, and field study therefore became a serious scientific discipline. The publication of *The Origin of Species* in 1859 had little immediate effect on the Victorian enthusiasm for natural history.[9]

In the early decades of the nineteenth century, natural history was synonymous with the study of animal, vegetable and mineral sciences – i.e. zoology, botany and geology – and the British Association was the major meeting ground. The term 'biology' (which had been used on the continent of Europe for half a century) was not introduced to England until 1862; it gradually replaced 'natural history' and assumed a broader base.

By the early nineteenth century, most museums contained heterogeneous assortments of curiosities devoid of systematic purpose or arrangement – anything from coins and corndollies to coleoptera! When the Ashmolean Museum was 'modernized' in 1820 it was arranged not on Linnaean lines but in accordance with Paley's *Natural Theology* – 'to induce a mental habit of associating the view of natural phenomena with the conviction that they are the media for Divine manifestation'. The British Museum's Natural History Department (Richard Owen, 1804–92, became Superintendent in 1856) was formed from collections bequeathed by Sir Hans Sloane (1668–1753). For a brief period at the end of the nineteenth century, the British Museum (Natural History) represented the fusion of scientific vitality and popular entertainment which characterized Victorian natural history; however, the idea was soon dismantled when Owen retired in 1884. A private collector, Walter (later Baron) Rothschild (1868–1937), opened a museum (which remains extant) containing a wide range of animals at Tring, Hertfordshire; this was a rival to the British Museum (which had been created by Act of Parliament in 1753).[10]

Richard Owen, a comparative anatomist and palaeontologist, assumed the role of the Frenchman Georges Cuvier (1769–1832) and continued the line of work begun by Linnaeus in the field of zoology; he was widely recognized to be the leading authority on all matters relating to zoological and palaeontological classification. However, his image as the 'Voice of Science' declined steeply when he failed to express a coherent view on Darwin's theory; the clear, witty and cogent remarks of T H Huxley won the day. To geologists, Roderick Murchison (1792–1871), who was involved in numerous societies and committees, was Owen's 'opposite number' in the field of natural history, Huxley's equivalent being Charles Lyell (1797–1875), who retained a low profile but was in retrospect the outstanding geologist of the era.[11]

By the end of the nineteenth century, preservation of the environment (a prominent theme today) was appreciated as being essential to the survival of all species; natural habitats must be preserved at all cost, and pollution prevented. Also, mere collecting was no longer acceptable; habitat, behaviour, diet, breeding habits and relation to other species were all considered important. A trend towards greater thoroughness was enhanced by the publication of *The Origin of Species*, which resulted from Darwin's detailed observations and research. From the 1880s, the microscope ceased to be a mere family treasure and became a standard piece of classroom equipment. By the end of the century, therefore, the conventional natural history textbook had become dull, and description of reactions to nature overtook the emphasis on memorizing long lists of Linnaean names.[12]

Manson's early career must have been profoundly influenced by natural history, which was such an extremely popular pursuit in Victorian England. Thus, he wrote (again in 1898):[13]

> It is evident ... that the student of medicine must be a *naturalist* [my italics] before he can hope to become a scientific epidemiologist, or pathologist, or a capable practitioner. The necessity for this ... is yearly becoming more apparent, but especially so in that section of medicine which relates to *tropical disease* [my italics].

And in the first Presidential Address to the Society of Tropical Medicine & Hygiene (STMH) in 1907, he considered:[14]

> It is apparent to any one who has followed the discoveries of recent years that *tropical medicine* [my italics] more than any other branch of medicine, is dependent on several of the collateral sciences, more especially on the various branches of *natural history* [my italics]. Our basis therefore has to be a broad one, and in drawing up the constitution of the [STMH] this point has been kept steadily in view.

Well into the twentieth century, several of the pioneers of the then well-established discipline of tropical medicine were also naturalists. G C Low (see Chapter 8) was an eminent ornithologist, and ornithology was also a major hobby of Sir Philip Manson-Bahr (see Chapter 18); both became Presidents of the (R)STMH.[15]

EVOLUTION: AN AWAKENING OF SCIENTIFIC INQUIRY

Evolutionary theories, of course, existed long before Charles Darwin (1809–82) – such theories included the descent theory, the continuity theory, the developmental hypothesis, and transmutation. In the 1840s Robert Chambers (1802–71) wrote *Vestiges of Creation* (1844), published anonymously, which attracted a great deal of attention. This book, which was frowned upon by most scientists, brought evolution to the public's attention, and in fact might have acted as a catalyst for Darwin's work. Naturalists had previously been unquestioning and had accepted that the age of the Earth was, as laid down by Archbishop Ussher (1581–1656), 4004 BC; everything in nature had been created by God for man's convenience. Paley had provided the theological philosophy, and Linnaeus the biological background. Popular natural history thus continued as before – in a vacuum! Darwin (like Wallace – see below) acknowledged only two major influences in the formulation of their theory: Lyell's *Principles of Geology* (published in three volumes between 1830 and 1833), and Thomas Malthus' (1766–1834) *An Essay on the Principle of Population, as it affects the Future Improvement of Society* (1798); neither was, of course, a biological work. Darwin claimed not to have been influenced by any (other) evolutionary theory – in particular that of Lamarck (see below).[16]

In *Histoire Naturelle*, Buffon (1707–88) surmised that the world was much older than had hitherto been considered possible. He alluded to many aspects of evolution, including hybridism, artificial breeding, rudimentary organs and the geographical distribution of species; however, although he believed in the mutability of species, he considered evolution to be a process of 'degeneration' from a perfect to an imperfect form, rather than a process of increasing specialization. The first clear formulation of an unequivocal theory of evolution was by Erasmus Darwin (1731–1802) – grandfather of Charles – in *Zoonomia: or the Laws of Organic Life* (1794–6). He argued that all species originated from the same primitive 'filament', and subsequently developed and differentiated under the influence of external forces.

Jean Baptiste Lamarck (1744–1829) published his evolutionary work, *Philosophie Zoologique*, in 1809, arguing that all species had evolved (and continued to do so) by a continuous progression, and that changes in structure arose in response to a 'new' environmental condition.

In 1831 Patrick Matthew published a theory of natural selection which, looked at retrospectively, is remarkably similar to that of Charles Darwin; unfortunately, he published it in an obscure text – *On Naval Travel and Arboriculture*.[17]

The development of evolutionary theory is coloured by several contrasting characters, all of whom warned their readers not to put too much reliance on natural theology. Hugh Miller (1802–56) attempted to justify recently acquired geological evidence in the context of the *Book of Genesis*; his best-known book, *The Old Red Sandstone*, became widely read. Philip Gosse (1810–88) adopted a fundamentalist approach, considering that the geological record was invented by the Creator to hoodwink man into believing that the Earth had a prior history, and furthermore that Genesis was after all correct in recording the date of the origin as 4004 BC.[18]

Lyell's book (which, like Darwin's, was accompanied by a wealth of supportive data) undoubtedly paved the way for Darwin's theory; this work undermined the literal acceptance of the *Book of Genesis* from a strictly geological viewpoint. His theory of Uniformitarianism (which incorporated the underlying assumption that the Earth is millions of years old) was published in 1830 and was contrary to previous theories, including the 'Catastrophist theory' enunciated by Cuvier, although it had in fact been proposed by the Scottish physician James Hutton (1726–97) in his *Theory of the Earth* (1788). Darwin's perception, that the Uniformitarianism theory could apply equally well to the organic as to the inorganic world, changed the course not only of natural history but also of science overall.[19]

Darwin was thus by no means the first 'evolutionist'. His claim to fame rests first on his concept of natural selection (or 'descent with modification'), and secondly on a painstaking accumulation of a vast body of factual evidence to support this. Darwin received help and support from many established scientists, including Sir William Hooker (1785–1865), JS Henslow (1796–1861), Asa Gray (1810–88), Joseph Hooker (1817–1911) and, most importantly, T H Huxley (1825–95). The simultaneous enunciation of natural selection by Alfred Russel Wallace (1823–1913) while exploring in Indonesia and Malaysia was probably also triggered by Lyell's and Malthus' texts, and also the *Vestiges*; his conclusions were produced in a very short time.

The seminal work was read to the Linnaean Society of London on 1 July 1858; however, neither Darwin nor Wallace was present. One reason why Wallace never achieved Darwin's fame is associated with his abiding interest in spiritualism and phrenology. The joint theory, and subsequent work, became a 'thorn in the flesh' of the Established Church; it was at the British Association's Oxford meeting in June 1860 that the Bishop of Oxford, Samuel Wilberforce (1805–73), led the well-known anti-Darwinian attack. *The Origin* therefore demolished the cosy tenets of natural theology (as laid down by Paley and others); far from being a benevolent and intelligent architect, natural selection was portrayed as being limitless, wasteful and cruel! For every single useful variation which nature might produce randomly, thousands of others were in fact doomed to perish.[20]

By the late nineteenth century, most educated individuals accepted evolution in some form; however, by no means all were convinced that the Darwin/Wallace theory was correct. It seems clear, though, that Manson was well convinced of the accuracy of the natural selection theory; *The Origin* had been published when he was only fifteen years old. In 1898, he was to write in the preface of his famous textbook:[21]

> Disease germs, their transmitting agencies, or their intermediate hosts, being living organisms, are, during their extra-corporal phases, necessarily competing organisms, and therefore liable to be preyed upon, or otherwise crushed out, by other organisms in the struggle for existence. The geographical range of such disease germs, therefore, will depend, not only on the prevalence of favourable conditions but, also, on the absence of unfavourable ones.

UNDERSTANDING OF DISEASE CAUSATION: DEVELOPMENT OF THE 'GERM THEORY'

Before the latter years of the nineteenth century, two opposing theories existed to explain the origin(s) and spread of plague and other infectious diseases: the 'miasmatists' considered that a noxious substance either existed in the atmosphere or arose from the earth during plague periods, while the 'contagionists' believed that healthy individuals developed the disease because they had been (directly or indirectly) in contact with a person suffering from the affliction. In practice, preventive measures (see Chapter 1) took both theories into account; the suggestion that rats were involved in disease transmission dates back to ancient writings.[22]

Until the Middle Ages, no official measures were in existence to counteract epidemics. The fundamental concept of isolation had been derived from the 'leper ritual' as described in the Bible. During the early centuries of the Christian era, leprosy was regarded as a contagious disease. In the early thirteenth century, a concept became accepted that disease (including plague, any febrile illness accompanied by a rash, granular conjunctivitis, 'the itch' and erysipelas) was dominated by contagion. The 'Black Death' (1347–8) influenced health administration in a way similar to that which cholera would do in Europe in the 1830s. In 1546 Girolamo Fracastoro (1483–1553) published *De contagione et contagiosis morbis* ('On contagion and contagious diseases'), distinguishing three forms of contagion: direct contact, indirect contact via fomites, and transmission from a distance. 'Minute bodies' (with the power of self-multiplication) were, he considered, all responsible. The French physician Guillaume de Baillon (1538–1616) had reintroduced a Hippocratic concept of an 'epidemic constitution' – later extended by the English physician, Thomas Sydenham (1624–89). Sydenham recorded in *Methodus curandi fibres* ('The method of treating fevers') (1666) that different (infectious) diseases were distinct entities. GM Lancisi (1654–1720), physician to the Pope, published a book entitled *De noxiis paludum effluvis* ('On the noxious effluvia of marshes') in 1717; it outlined the miasmatic theory, although the author believed that malaria was in fact transmitted and, furthermore, that the mosquito was probably involved![23]

For two centuries following Sydenham, any discussion on the mode of spread of epidemics was futile; the miasmatists and the contagionists held widely opposing viewpoints. In practice, however, measures adopted to control infectious diseases were usually based on 'contagion' as being the cause. In 1840, Jakob Henle (1809–85) published *Von den Miasmen und Kontagien* ('On miasmata and contagion'); he divided epidemic diseases into three groups:

1. Miasmas (of which malaria was the prime example)
2. Those originating in miasmas, but from which a living parasite developed and proliferated within the body and conveyed the disease to others (most epidemics were included in this group)
3. Those involving contagion (this group included syphilis and scabies).

Henle was perhaps the first to clearly lay down the principles for a specific origin of infectious disease(s). At the close of the nineteenth century, when the organisms associated with several infections (and subsequently demonstrated to be causative) and the lifecycles of several parasites had been delineated, even some contagionists sought refuge in miasmatic theory.[24]

In the mid-nineteenth century, the views of Arnott and Southwood-Smith (an authority on 'fevers', and author of a *Treatise on Fever*, 1830), as well as many others were that 'fever' was caused by pythogens. The 'pythogenic theory' held that fevers were caused by an unknown atmospheric influence, only present at the time of epidemics, which acted on the noxious exhalations emanating from putrid animal or vegetable matter in ditches, stagnant drains and cesspools, and within the houses themselves. (This in essence was the miasmatic theory.) When a case of disease occurred, testing the house drains was therefore the first logical administrative procedure; the belief that diphtheria was caused by 'bad smells' died hard, and in the cases of cholera and typhoid a miasmatic concept was virtually universal. Southwood-Smith discounted the contagion theory completely; as a result, Chadwick virtually discarded quarantine, a fact which underlay a recommendation in his *Sanitary Report* that attention to sewers (which should be flushed with water) was of paramount importance in disease prevention.[25]

The 'germ theory' of infection

Several factors were involved in overturning the miasmatic theory:

- The discovery of bacteria
- Proof that bacteria are causatively related to disease
- The demonstration that bacteria arise in patients suffering from these diseases, and
- Perhaps most importantly, abandonment of the doctrine of 'spontaneous generation' (see below).

Anthonie van Leeuwenhoek (1632–1723) had first visualized living organisms (protozoa) in 1683; bacteria were next described by the Dane O F Müller (1730–84), who wrote *Animaleuta infusoria fluviatilia et marina* ('River and marine animalcules'), published posthumously in 1786. Demonstration that bacteria cause specific diseases originated with an observation by the Italian A Bassi (1773–1856) that an epidemic disease of silkworms was caused by a minute fungus *Botrytis bassiana*; this work was published in 1835–6. In 1839, the German physician J L Schölein demonstrated that another fungus, *Achorion schönleinii*, caused a dermatological disease, *flavus*. In the following year, J Henle published his famous essay on miasmas and contagion. Meanwhile, a significant advance in helminthology was due to George Busk (1807–86), a surgeon working on the second of the Seamen's Hospital ships, moored off Greenwich, who first demonstrated a large fluke (later named *Fasciolopsis buski*) in the intestine of a lascar.[26]

A belief that living organisms (even ones of considerable size) can be generated spontaneously from decaying matter dates at least to classical times. As early as the seventeenth century, Francesco Redi (1626–97) showed at Arezzo that maggots in decaying flesh do not in fact arise spontaneously, but emerge from eggs laid by flies; he thus challenged 'spontaneous generation'. Between 1765 and 1776 Spellanzani demonstrated that if an infusion of hay is boiled and the air in contact with it heated, with complete exclusion of external air, no animalcules develop; moreover, a higher order of organism was easily destroyed by heat, and a lower one (probably including bacteria) resisted boiling for 30 minutes. These experiments should have proved conclusive, but controversy continued, and the subject in the mid-nineteenth century remained confused. Further controversy surrounded the question of fermentation – did it take place with or without microbes, i.e. was it purely a chemical process?[27]

Louis Pasteur (1822–95; Figure 2.1), an outstanding French chemist, solved numerous problems. He worked on 'ferments' – living cells that exert a putrefactive action; these, he believed, were conveyed by atmospheric dust to wort or beer. It was at that time widely held that fermentation, putrefaction and infection had much in common. Pasteur's opponents argued, however, that the living organisms present during fermentation were the *result* of the process (spontaneous generation) and not the cause. He established that air containing living organisms was capable of producing putrefaction; the organisms lost their power, however (presumably being killed), by heating. He published these and other experiments in *Mémoire sur les corpuscles organises que existent dans l'atmosphère* ('Memoir on the organized bodies which exist in the air') in 1861.

The first disease studied by Pasteur which affected *Homo sapiens* was anthrax. He demonstrated that under certain conditions, bacilli produce highly heat-resistant spores. In an outstanding series of experiments the German microbiologist

FIGURE 2.1 Louis Pasteur (1822–95), a French chemist who launched the 'germ theory' of disease (reproduced courtesy of the Wellcome Library, London).

Robert Koch (1843–1910; Figure 2.2) carried out detailed studies on anthrax, while Pasteur continued work on this disease as well as on rabies and fowl cholera. Koch succeeded in establishing bacteriology as a distinct science; in 1882 he demonstrated *Mycobaterium tuberculosis*, the causative agent of tuberculosis, and later worked on cholera and African trypanosomiasis. The English surgeon Joseph (later Lord) Lister (1827–1912; Figure 2.3) published his first paper on antiseptic surgery in 1867, thus clinching the 'germ theory'. Use of this antiseptic method produced a fundamental revolution in surgery. Council minutes of a meeting of the (Royal) Society of Tropical Medicine and Hygiene held on 16 February 1912 record that:[28]

> The following resolution … prepared by the President [Sir William Leishman; see Chapter 12] [was] carried unanimously:- The Society of Tropical Medicine and Hygiene, meeting on the day of the funeral of the late Lord Lister, desires to place on record an expression of profound respect for the memory of one, who, in his generation, has contributed more

FIGURE 2.2   Robert Koch (1843–1910) established the science of bacteriology (reproduced courtesy of the Wellcome Library, London).

to the progress of science and its application to the benefit of mankind than any other individual. The Society recognizes that the practical connection traced by Lord Lister between
micro-organisms and diseases, laid the foundation for the work which is now being done
by this and kindred Societies in this same direction.

Snow's observation in the mid-nineteenth century (see Chapter 1) had demonstrated indirectly that cholera resulted from contaminated drinking-water, and
this swept away the view that this disease was airborne. In a similar way, Ross
had in 1897 dispelled the view that malaria was directly caused by miasmas
derived from marshes.

Now, the responsible organisms themselves had been demonstrated. Manson,
writing in 1898, was clearly committed to the 'germ theory' despite the fact that at
the time the miasmatic theory of disease was still held by many:[29]

Modern science has clearly shown that nearly all diseases, directly or indirectly, are caused
by germs.

FIGURE 2.3   Joseph (later Lord) Lister (1827–1912) first used the antiseptic method in surgery
(reproduced courtesy of the Wellcome Library, London).

CLINICAL PARASITOLOGY

Some writers trace the origin of tropical medicine, as a discipline, to Manson's seminal discovery in 1877 – i.e. the demonstration that microfilariae of *Wuchereria bancrofti* are transmitted to the mosquito from man. Others have given the priority to his three classical Goulstonian Lectures, delivered to the Royal College of Physicians in 1896 (see Chapter 3), in which he outlined the life-cycle of *Plasmodium* spp. (before Ross's work in India). Both undoubtedly proved instrumental in launching the new specialty. The sum total of these momentous discoveries was that mosquito transmission of a nematode (lymphatic filariasis), protozoan (*Plasmodium* spp.), and viral (yellow fever) disease were all established within the course of little more than two decades. In 1898, Manson wrote:

> in the study of tropical disease bacteriology may be relegated to quite a secondary place. The student's attention is confined almost entirely to protozoa and helminths, to the special vectors or media of these organisms, to their pathological effects, and to the prophylaxis and treatment of the diseases they give rise to.

However, despite these suggestions there can be no doubt that the specialty of *tropical medicine* owes its origin to a multiplicity of unrelated disciplines, as outlined above. Furthermore, it seems clear that Manson himself, the 'father of modern tropical medicine', duly acknowledged this fact![30]

NOTES

1 P Manson. *Tropical Diseases: a manual of the diseases of warm climates.* London, 1898: Cassell and Co Ltd; J Morris. *Pax Britannica: The Climax of an Empire.* London, 1979: Penguin Books, p. 544; G C Cook. *From the Greenwich Hulks to Old St Pancras: a history of tropical disease in London.* London, 1992: Athlone Press, p. 338; G C Cook. Evolution: the art of survival. *Trans R Soc Trop Med Hyg* 1994, 88: 4–18; T Pakenham. *The Scramble for Africa 1876–1912.* London, 1991: Weidenfeld and Nicolson, p. 738.

2 C Singer, E A Underwood. *A Short History of Medicine,* 2nd edn. Oxford, 1962: Clarendon Press, p. 854; A S Wohl. *Endangered Lives: public health in Victorian Britain.* London, 1983: Methuen & Co Ltd, p. 440; G Himmelfarb. *The Idea of Poverty: England in the early Industrial Age.* London, 1984: Faber and Faber, p. 595; R Trench, E Hillman. *London Under London: a Subterranean Guide.* London, 1984: John Murray, p. 224.

3 *Op cit.* See note 2 above. See also C Wilcocks. A historical trend in tropical medicine. *Trans R Soc Trop Med Hyg* 1963, 57: 395–408; G C Cook. Thomas Southwood Smith FRCP (1788–1861): leading exponent of diseases of poverty and pioneer of sanitary reform in the mid-nineteenth century. *J Med Biog* 2000, 10: 194–205.

4 G Weightman. *London River: the Thames story.* London, 1990: Collins and Brown Ltd, p. 160; G C Cook. Joseph William Bazalgette (1819–1891): a major figure in the health improvements of Victorian London. *J Med Biog* 1999, 7: 17–24; G C Cook. Construction of London's Victorian sewers: the vital role of Joseph Bazalgette. *Postgrad Med J* 2001, 77: 802–4. See also *op cit.* note 2 above (Singer, Underwood; Trench; Hillman).

5 W J R Simpson. Some considerations regarding preventable diseases and their prevention. *Trans R Soc Trop Med Hyg* 1919, 13: 31–44; H H Scott. *History of Tropical Medicine: based on the Fitzpatrick lectures delivered before the Royal College of Physicians of London 1937–38,* two volumes. London, 1939: Edward Arnold, p. 1165.

6 W MacGregor. The problems of tropical medicine. *Lancet* 1900, ii: 1055–61; W Osler. The nation and the tropics. *Lancet* 1909, ii: 1401–6; A Balfour. The problem of hygiene in Egypt. *Lancet* 1919, ii: 417–21, 467–70, 507–12; A Balfour. An address on why hygiene pays. *Br Med J* 1926, i: 929–32. See also P Manson. *Tropical Diseases: a manual of the diseases of warm climates.* London, 1898: Bailliére Tindall; R Ross. The future of tropical medicine. *Trans Soc Trop Med Hyg* 1909, 2: 272–85.

7 G C Cook. Doctor David Livingstone, FRS (1813–1873): 'the fever' and other medical problems of mid-nineteenth century Africa. *J Med Biog* 1994, 2: 33–43; C Hibbert. *Africa Explored: Europeans in the Dark Continent, 1769–1889.* London, 1982: Allen Lane, p. 336; D Middleton. *Francis Galton: travel and geography.* In: M. Keynes (ed.), *Sir Francis Galton, FRS: the legacy of his ideas.* London, 1993: Macmillan Press, p. 237; J Blackburn. *Charles Waterton: traveller and conservationist 1782–1865.* London, 1989: Century, p. 243.

8 L Barber. *The Heyday of Natural History 1820–1870.* New York, 1980: Doubleday & Co Inc, p. 320; G White. *The Natural History of Selborne* (ed. with notes by E Blyth). London, 1836, p. 418; R Mabey. *Gilbert White: a biography of the author of the Natural History of Selborne.* London, 2006: Profile Books, p. 239; H B Carter. *Sir Joseph Banks 1743–1820.* London, 1988: British Museum (Natural History), p. 671; C Linnaeus. *Systema naturae, sive regna tria naturae systematiae proposita per classes, ordines, genera, & species.* Sweden, 1735: Theodoum Haak; W Paley. *Natural Theology, or Evidences of the Existence and Attributes of the Deity.* London, 1802; C Darwin. *On the Origin of Species by Means of Natural Selection, or the preservation of favoured races in the struggle for life.* London, 1859: John Murray; L Tree. *The Ruling Passion of John Gould: a biography of the bird man.* London, 1991: Barrie and Jenkins, p. 250.

9 *Op cit.* See note 8 above (Barber, Darwin).

10 *Op cit.* See note 8 above (Barber, Paley).

11 Sir Richard Owen (1804–92) was a prominent comparative anatomist and palaeontologist who was essentially an opponent of Charles Darwin's theory of evolution. See also N A Rupke. *Richard Owen: Victorian Naturalist.* London, 1994: Yale University Press, p. 462; J W Gruber. Owen, Sir Richard (1804–92). In: H C G Matthew, B Harrison (eds), *Oxford Dictionary of National Biography*, Vol. 42. Oxford, 2004: Oxford University Press, pp. 245–54. *Op cit.* See note 8 above (Barber).

12 *Op cit.* See note 8 above (Darwin).

13 *Op cit.* See note 6 above (Manson).

14 P Manson. Inaugural address. *Trans Soc Trop Med Hyg* 1907, 1: 1–12.

15 G C Cook. George Carmichael Low FRCP: twelfth President of the Society and underrated pioneer of Tropical Medicine. *Trans R Soc Trop Med Hyg* 1993, 87: 355–360; G C Cook. Correspondence from George Carmichael Low to Dr Patrick Manson during the first Ugandan sleeping sickness expedition. *J Med Biog* 1993, 1: 215–29.

16 D Young. *The Discovery of Evolution.* London, 1992: Natural History Museum Publications, p. 256; F Darwin (ed.). *The Autobiography of Charles Darwin and Selected Letters.* New York, 1958: Dover Publications Inc, p. 365; E J M Bowlby. *Charles Darwin: A Biography.* London, 1990: Hutchinson, p. 511. P J Bowler. *Charles Darwin: the man and his influence.* Oxford, 1990: Basil Blackwell, p. 250; A Desmond, J Moore. *Darwin.* London, 1991: Michael Joseph, p. 808. *Op cit.* See note 8 above (Barber, Paley).

17 *Op cit.* See note 8 above (Barber).

18 *Ibid.*

19 *Ibid.*

20 A R Wallace. *The Geographical Distribution of Animals with a Study of the Relations of Living and Extinct Faunas as Elucidating the Past Changes of the Earth's Surface.* London, 1876: MacMillan and Co. See also *op cit.* notes 8 (Barber), 16 (Young), 19 (Desmond, Moore) above.

21 *Op cit.* See note 6 above (Manson).

22 *Op cit.* See note 2 above (Singer, Underwood).

23 *Ibid.* See also *op cit.* note 1 above (Cook).

24 *Op cit.* See note 2 above (Singer, Underwood).

25 *Op cit*. See note 3 above (Cook).
26 *Op cit*. See notes 1 (Cook) and 2 (Singer, Underwood) above.
27 *Ibid* (Singer, Underwood).
28 Society of Tropical Medicine and Hygiene: Council Minutes: 1912: 16 February.
29 *Op cit*. See note 6 above (Manson).
30 *Ibid*. See also Anonymous. Lymphatic filariasis – tropical medicine's origin will not go away. *Lancet* 1987, i: 1409–10.

4

Alphonse Laveran (1845–1922): discovery of the causative agent of malaria in 1880

This chapter is centred on the discovery of the causative agent of malaria; the description and controversy surrounding the *transmission* of the disease is covered in Chapter 5. Miasma still dominated contemporary thought at this time, with malaria being the classical disease caused by this means.

Charles Louis Alphonse Laveran (1845–1922; Figure 4.1) was a Frenchman, born in Paris. His father was a Professor at the Military Hospital of Metz, but he went to Algeria with his family in 1850; six years later the family returned to France. Alphonse decided to follow in his father's footsteps, and joined the French Army. He thenceforth proceeded to the Collège Sainte-Barbe and the Lycée Louis-le-Grand, where he took his baccalaureate. Laveran spent the next four years at the École de Service de Santé Militaire, where in 1866 he successfully defended his doctoral thesis, *Regeneration of Nerves*. He then proceeded to the École du Val-de-Grâce, after which he became aid-major at the military Hôpital Saint-Martin. When the Franco-Prussian War (1870–71) started, Laveran was posted to the ambulances of the eastern army. He was later involved in the battle of Gravelotte and Saint-Privat, and the Siege of Metz. In 1870 he returned to the military hospital at Lille, and the following year to the Hôpital Saint-Martin.

FIGURE 4.1   Alphonse Laveran (1845–1922), discoverer of the cause of *Plasmodium* spp. infection in 1880 (reproduced courtesy of The Wellcome Library, London).

By 1874 Laveran was accredited as Professor of the Diseases and Epidemics of Armies at Val-de-Grâce, and in 1878 he was posted to Constantine, Algeria. When he started work in Algeria he recognized that the high mortality rate in soldiers was largely a result of *fièvres palustres*, which he, like most of his contemporaries, initially termed a 'telluric' disease, reflecting the current view that the disease arose from the effluvia of wet or marshy soil (see Chapter 1). He began at the military hospital at Bône, where he carried out post-mortems and examination of the blood of malaria 'victims'. It was in late 1880 that Laveran made his seminal discovery (see below, also Figure 4.2).

Laveran also established the Laboratory of Tropical Diseases at the Pasteur Institute (1907), and the Société de Pathologie Exotique (1908). He wrote extensively, and his works include *Traité des maladies et epidemics des armies* (1875), *Traité des fièvres avec la description des microbes du paludisme* (1894), and *Trypanosomes et trypanosomiasis* (1904).

FIGURE 4.2 Stamp commemorating the centenary, in 1980, of Laveran's discovery (reproduced courtesy of The Wellcome Library, London).

He later wrote of his important discovery:[1]

> In 1880, as I was trying to account for the mode of formation of the pigment in the palustral blood, I was led to see that besides melaniferous leucocytes, spherical hyaline corpuscles without nucleus could be seen, and also very characteristic crescent-shaped bodies.
>
> I had proceeded thus far with my researches, and was still hesitating whether these elements were parasites, when on 6 November 1880, on examining the pigmented spherical bodies mentioned above, I observed, on the edge of several of these elements, moveable filaments or flagella, whose extremely rapid and varied movements left no doubt as to their nature.

What Laveran had so accurately described was ex-flagellation from the male gametocyte ('crescent'), which he was able to witness clearly because the blood was fresh; he could not possibly have made this important observation had the blood been 'fixed' and/or stained.

Laveran continued, however, with his belief that malaria arose from marshes, until Ross's work in 1897/8 (see Chapter 5). He also wrote:[2]

> These parasites evidently live at the expense of the blood corpuscles, which grow pale in proportion as the parasitic elements which adhere to them increase in size: there is a time when the blood corpuscle is only distinguished by its contour, its characteristic colour has disappeared.

Laveran surmised that the 'parasites ... adhere to the [erythrocytes] by pressing upon them'.

Later, he carried out a great deal of research on leishmaniasis, trypanosomiasis and other protozoan infections, and was thus a powerful influence on the subsequent development of tropical medical research. He joined the Pasteur Institute, Paris in 1897, and received the Nobel Prize for Physiology or Medicine, although not until 1907.[3] He explained the occurrence of paludism in non-marshy areas as being a result of contaminated drinking water, which should always be boiled.

LAVERAN'S 1880 DISCOVERY: THE TRUE *CAUSE* OF MALARIA IS REVEALED

Widespread acceptance of the causative agent of malaria (now that the 'germ theory' was taking off) was a very lengthy saga. Laveran's seminal discovery had taken place on 6 November 1880 in Constantine, Algeria, and consisted of finding male gametocytes (crescents) of *P. falciparium* in a blood sample of a malarious soldier. Laveran maintained from the outset that there was only one species of malaria parasite.

Laveran named his newly found organism *Oscillaria malariae*, and this finding was communicated to the Academy of Medicine in Paris on 23 November and 28 December 1880. On 12 November 1881, the work was published in *The Lancet*. Claims for priority were immediately made – on behalf of Philippe Klencke (1813–81) in 1843, and Maxime Cornu (1843–1901) in 1871. Laveran was probably *not* the first investigator to visualize the malaria parasite. Eight years previously, in 1872, Delafield had published a small volume, but did not appreciate the significance of the bodies which he saw.

Laveran's demonstration had been preceded by a series of false attempts. Thus, JK Mitchell (1793–1858), working in Philadelphia, apparently claimed that a fungus was the cause, and in 1879 Edwin Klebs (1834–1913) and Corrado Tommasi-Crudeli (1834–1900) were convinced that a bacterium, *Bacillus malariae*, was responsible; this was supported by Ettore Marchiafava (1847–1935; Figure 4.3), Guiseppe Cuboni (1852–1920) and Eduardo Perroncito (1847–1936). However, G M Sternberg (1858–1915) failed to confirm the observation. As late as 1898, Dunley-Owen (a surgeon to the British South Africa Company), working in Rhodesia, demonstrated that horse sickness (considered at that time to be a form of malaria) only occurred in animals turned out at or before sunrise – evidence that the malaria poison (a bacterium) was only virulent for one hour before and soon after sunrise.[4]

Doubts about Laveran's malarial parasite

There was initially a great deal of scepticism about Laveran's malarial parasite. It took a great deal of persuasion to convince peers that *B. malariae* was *not* involved. Marchiafava (see above) and Angelo Celli (1857–1914; Figure 4.4), in assessing Laveran's work, disagreed with his interpretation but claimed that he had (i) drawn attention to the non-leucocytic nature of the pigment-containing granules, and (ii) that these granules were contained in erythrocytes. They considered that Laveran's 'bodies' were degenerate erythrocytes – the products of regressive metamorphosis brought about by the bacterium *B. malariae*. However, Camillo Golgi (1844–1925; Figure 4.5) was convinced of the value of Laveran's observations; he furthermore showed that the cyclical development of the 'parasite' coincided with the rigors – so prominent a clinical feature of the disease. Golgi later described *P. malariae* and indicated that quartan and tertian fevers were caused by distinct species; he considered that in the benign forms the

FIGURE 4.3   Ettore Marchiafava (1847–1935), founder of the Italian School of malariology (reproduced courtesy of The Wellcome Library, London).

development of the parasite took place in the circulating blood, while in the 'pernicious' or malignant variety it occurred within organs – the brain included.

Meanwhile, Robert Koch (1843–1910) remained unconvinced by Laveran's observations, but he was warmly welcomed at the Pasteur Institute! In 1885, William Councilman (1854–1933) and Alexander Abbott (1860–1935), working at Baltimore, also remained sceptical of Laveran's findings; however, another American, G M Sternberg (1858–1915) came to Laveran's defence, maintaining that 'the peculiar blood parasite [described by Laveran] is directly concerned in the aetiology of the malarial fevers'.[5] Later Councilman changed his mind, and was followed by William Osler (1849–1919; Figure 4.6), who wrote in 1886, in an editorial entitled *The malarial germ of Laveran* (see Figure 4.7):[6]

> The description is that of a flagellatic infusorian with an amoeboid phase of development, and should this body turn out to be the veritable germ of malaria, it will be the first instance of a wide-spread endemic affection associated with protozoan forms.

FIGURE 4.4   Angelo Celli (1857–1914) (reproduced courtesy of The Wellcome Library, London).

Osler was therefore now 'converted', and his change of heart was presented to the Pathological Society of Philadelphia in 1886. He became convinced of their aetiological association with malaria, citing (i) the destruction of erythocytes (which is one of the most marked symptoms), and (ii) the specific action of quinine on the parasites.

Continuing scepticism

Osler was himself an experienced microscopist and a pioneer amongst haematologists of his era. Under an entry for June 1886, his biographer, Cushing, wrote:

> In the very first [case to be examined at Blockley, Philadelphia], on April 20th, he [Osler] had unquestionably seen and made drawings of the amoeboid stage of the malarial parasites, though he was evidently uncertain as to their interpretation … it was the beginning of his great interest in malaria, but it was not until his studies were resumed in the autumn that he became fully convinced of the truth of Laveran's claims regarding the protozoal origin of the disease.[7]

FIGURE 4.5 Camillo Golgi (1844–1925) (reproduced courtesy of The Wellcome Library, London).

and in January of the following year:[8]

> with the beginning of the year he had resumed his studies of malaria with an attempt to differentiate the organisms of various types of the disease – quotidian, tertian, and quartan, anticipating the observations of Marchiafava and Celli published two years later.

These early researches were in fact summarized by Osler in an address delivered to the Association of American Physicians, and also to the Pathological Society of Philadelphia on 28 October 1886.[9] Following an historical overview of Laveran's researches, as well as some technical detail, he proceeded to a series of accurate descriptions (accompanied by sketches; see Figure 4.7) of various stages in the evolution of *Plasmodium malariae*, with a diagram of parasitized

FIGURE 4.6   William Osler (1849–1919), who shared early scepticism regarding Laveran's finding (reproduced courtesy of The Wellcome Library, London).

erthrocytes in a cerebral capillary. However, like many workers of that time, he remained sceptical that these 'organisms' were in fact *causative*:[10]

> The same difficulty meets us here as in so many affections in which micro-organisms have been found: Are they pathogenic, or are they merely *associated* [my italics] with the disease, which in some way furnishes conditions favourable to their growth?

Overall, however, Osler seems to have favoured a causative role:[11]

> First, the positive anatomical changes which can be directly traced to their action … the destruction of the red blood-corpuscles … The second fact is the action of quinine upon the parasites. The simultaneous disappearance of the symptoms of the disease and the haematozoa suggest that the specific influence of the medicine is upon the parasites, though it may be urged that the quinine, while curing the disease, simply removes the conditions which permit of their growth in the blood.

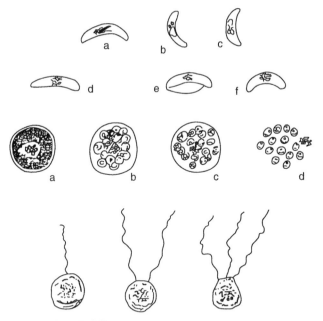

FIGURE 4.7   Osler's sketches of Laveran's organism.

Following this, he seems to have become less interested in malaria, until Laveran's observation was widely accepted.[12] Ross was also highly sceptical of Laveran's discovery until, in 1894, he had the malaria parasite demonstrated to him by Manson (see Chapter 5).

The *Lancet* proclaimed in an anonymous annotation in 1888 that:[13]

> Such a concurrence of testimony from three widely separated corners of the globe [Laveran, Marchiafava and Celli, Osler and [Henry Vandyke] Carter (1831–97)] is so striking as to serve to establish the discovery as fact, although, of course, it would be premature to assert that the organism in question is the *sole* [my italics] agent of the disease.

In 1891, scepticism remained. J Mayer wrote in *The New Orleans Medical and Surgical Journal*:[14]

> Whether the malaria is due to the protozoa described by Laveran, whose observation had so quickly been confirmed by Golgi, Marchiafava and Celli in Italy, Carter in India, and Councilman, Osler and James in this country, is of course still an open question.

Marchiafava and the other sceptical Italian malariologists slowly became convinced, however, of Laveran's observations between 1882 and 1885, and in 1886 gave the name *Plasmodium* to the genus.

Manson becomes convinced

In December 1893 Manson gave a lecture at University College Hospital, London, in which he gave a demonstration of Laveran's findings,[15] followed, in 1894, by observations on the significance of these findings.[16] This was, of course, *before* Manson (or anyone else) hypothesized what was going on outside the human body although, in the light of his observations on lymphatic filariasis, Manson must already have had a shrewd suspicion that the mosquito was involved.

Gradual acceptance of Laveran's observation owed a great deal to improvements in staining techniques; both Laveran and Osler were working with unstained films.

RETROSPECTIVE VIEWS OF P C C GARNHAM

Cyril Garnham (1901–94) was certainly not a *pioneer* of the discipline, by definition. He was, however, an expert on malarial research and, together with Shortt, delineated the exo-erythrytic cycle of *P. vivax* infection (see Chapter 15). No-one knew the malaria literature better than he.

Garnham devoted his Presidential Address to the Royal Society of Tropical Medicine and Hygiene in 1976 to an analysis of this work. In his judgement, the day on which 'Laveran first recognized the exflagellating body, [was] perhaps the most important day for tropical medicine', and he decided to pay tribute to five of the thirteen Honorary Fellows (four of whom worked in 'remote outposts') who were elected in 1907 (the year the Society was inaugurated – see Chapter 7). Rome, owing to its long history of marshes and malaria, was, he claimed, the 'Mecca of the malariologist'.

Turning to Marchiafava, he became convinced of Laveran's claims (although he had had serious doubts) despite the fact that he worked for the bacteriologist Tommasi-Crudeli (see above), who held that *B. malaria* (a bacterium) was the causative agent of malaria. Marchiafava and Celli named Laveran's parasite *Plasmodium malariae*. By injecting blood containing *P. falciparum* into another individual and producing a new infection, these workers were the first to prove the parasitic nature of aestivo-autumnal fever (malignant tertian malaria). Marchiafava formulated theories as to the *cause* of cerebral malaria – mechanical alteration of the circulation, and the probable involvement of toxins.

It was Marchiafava who, in Garnham's view, founded the Rome School of malariologists; his colleagues included: Bignami (see Figure 4.8), Celli, Bastianelli (see Figure 4.9), Dionisi and Sanfelice (see Chapter 5). Although the Italian School was centred on Rome, Golgi, a 'lone worker', was based at Pavia in northern Italy – where, a century earlier, Lazaro Spallanzani (1729–99) had been employed. His improved staining technique led to advances in the fine anatomy of the nervous system; he had described the Golgi apparatus (body). Unlike Laveran, who held that there was only one malarial parasite, Golgi demonstrated different parasites causing quartan and tertian fevers by showing different periodicity of the parasite cycle in the blood.

P. falciparum was described and named by William Welch (1850–1934) in 1887, but *P. ovale* was not described until 1922, by J W W Stephens (1865–1946).

FIGURE 4.8   Amico Bignami (1862–1929) (reproduced courtesy of The Wellcome Library, London).

Elucidation of the complete lifecycles of the four human *Plasmodium* species took, of course, nearly a century to complete.

Garnham also spoke in his lecture of Koch's (the 'father of bacteriology') contributions to tropical medicine (see Chapter 2). He initially 'dismissed the structures [demonstrated to him by Marchiafava] as artefacts', but later admitted his mistake. Koch in fact worked in East Africa, Java and New Guinea – where he studied hyperendemic malaria.

The last pioneer to be included in Garnham's evaluation was Vasili Danilewsky (1852–1939), from the Ukraine. His major contribution to malariology was recognition that avian *Haemosporidia* were similar to the human malaria parasite as described by Laveran; he was therefore 'well aware of the signifi-cance of bird malaria in the interpretation of human malaria'. This contribution

FIGURE 4.9   Guiseppe Bastianelli (1862–1959) (reproduced courtesy of The Wellcome Library, London).

was obviously of extreme importance when we consider the *transmission* of malaria (see Chapter 5).[17]

CONCLUSION

The seminal discovery of the causative agent of malaria, which altered views dating back to time immemorial, was thus made by the Frenchman, Alphonse Laveran, on 6 November 1880. His discovery faced a great deal of scepticism before being finally accepted. The ground had thus been prepared for elucidation of the mode of *transmission* of the infection to *Homo sapiens* (see Chapter 5).

NOTES

1 A Laveran. *Paludism* (trans. J W Martin). London, 1893: New Sydenham Society, p. 197.

2 A Laveran. Deuxième note relative à un nouveau parasite trouvé dans le sang des maladies attents de fiêve palustres: origine parasitaire des accidents de l'impaludism. *Bull de l'acad de méd* 1880: 9.

3 M Phisalix. *Alphonse Laveran: sa vie, son oeuvre.* Paris, 1923: Masson, p. 268. E R Nye. Alphonse Laveran (1845–1922): discoverer of the malarial parasite and Nobel laureate, 1907. *J Med Biog* 2002, 10: 81–7.

4 A Laveran. The pathology of malaria. *Lancet* 1881, ii: 840–41; A Delafield. *Handbook of Post-mortem Examinations and of Morbid Anatomy.* New York, 1872: William Wood & Co; A Dunley-Owen. Some notes on malaria as seen in Rhodesia. *Lancet* 1898, ii: 798–9; *Ibid.* The 'blind-fly' and the locust in the evolution of the malaria parasite. *Lancet* 1898, ii: 1764–5; R S Desowitz. *The Malaria Capers: more tales of parasites and people, research and reality.* London, 1991: W W Norton & Co, p. 288.

5 C M Poser, G W Bruyn. From swamp fever to the malarial parasite. In: *An Illustrated History of Malaria.* New York, 1999: The Parthenon Group, pp. 21–35.

6 W Osler. The malarial germ of Laveran. *Med News* 1886, 49: 265–6.

7 H Cushing. *The Life of Sir William Osler.* London, 1940: Oxford University Press, p. 1417.

8 *Ibid.*

9 W Osler. The haematozoa of malaria. *Trans Pathol Soc Philadelphia* 1887, 13: 255–76.

10 W Osler. An address on the haematozoa of malaria. *Br Med J* 1887, i: 556–62.

11 *Ibid.*

12 G C Cook. William Osler's fascination with diseases of warm climates. *J Med Biog* 1995, 3: 20–9.

13 Anonymous. *Lancet* 1888, i: 1201–2.

14 F Mayer. A few clinical notes. *New Orleans Med Surg J* 1890/1891, 18: 351–61.

15 P Manson. A clinical lecture on the parasite of malaria and its demonstration. Delivered at the University College Hospital on Dec 15th 1893. *Lancet* 1894, i: 6–9.

16 P Manson. On the nature and significance of the crescentic and flagellated bodies in malarial blood. *Br Med J* 1894, ii: 1306–8.

17 P C C Garnham. Reflections on Laveran, Marchiafava, Golgi, Koch and Danilewsky after sixty years. *Trans R Soc Trop Med Hyg* 1967, 61: 753–64.

5

Ronald Ross (1857–1932): the role of the Italian malariologists, and scientific verification of mosquito transmission of malaria

The supposition that malaria arose from marshes was by no means extinguished by Laveran's discovery of the causative organism. Indeed, Laveran himself, in company with most at the time, still held that man acquired the infection via contaminated drinking water. For several centuries there had been a suspicion that the mosquito was in some way involved (see Chapter 1 and below), but that if this were the case, man probably acquired the disease via the gastrointestinal tract – i.e. by drinking water contaminated by dead mosquitoes.

THE SCENARIO BEFORE ROSS

The suspicion that malaria transmission is caused by an insect is thus very old indeed and has been well reviewed by Poser and Bruyn. Since early Roman times a connection between mosquitoes and malaria was a widely held view, and this was entertained by Giovanni Lancisi (1654–1720) in 1717, although he combined it with the miasma theory. Benjamin Rush (1745–1813) wrote in 1790, however, that it seemed impossible that 'fevers' could be carried from marshes by insects. L D Beauperthuy (1807–71), a French doctor practising in Venezuela, held in 1854–6 that mosquitoes were involved in transmitting the 'intermittent fever'. And Giovanni

Rasori (1766–1837) reiterated his belief that an insect was causative in malaria transmission. Probably the first to clearly implicate the mosquito was, however, J C Nott (1804–73) in 1848.[1]

It seems probable, therefore, that the inhabitants of several malarious regions of the tropics had for many centuries associated malaria transmission with mosquitoes, but this association had never been confirmed in either lay or scientific literature.[2]

In 1882 A F A King (1841–1914) firmly subscribed to the mosquito hypothesis, reading a paper entitled 'Insects and disease – mosquitoes and malaria' to the Philosophical Society of Washington in February of that year, sixteen months after Laveran's discovery (see Chapter 4). He marshalled no less than nineteen facts in support of his theory, but regrettably he published this in a lay journal. King was born in England, his father being a physician in Oxfordshire. Moving to the USA at ten years old, he graduated from the National Medical College (NMC) and the University of Pennsylvania in 1865. King became Professor of Obstetrics at the NMC and the University of Vermont.[3]

Laveran (see Chapter 4) speculated in 1883 that mosquitoes were involved in malaria transmission, and in 1889 Carl Flügge (1847–1923) also considered that, in addition to air and water, mosquitoes were involved. In 1892, Richard Pfeiffer (1833–1902) cited animal experiments to support this theory. In 1896, Francesco Mendini (d. 1903) also supported mosquito involvement.[4]

The idea that a parasite could be passed from one host to a different species (metataxy) had been conceived by P C Abildgaard (1740–1801) in Holland in 1790. Almost a century later, in 1869, Alexei Fedchenko (1844–73) demonstrated that *Cyclops* was involved in the lifecycle of *D. medinensis*. Soon afterwards, Manson showed that *W. bancrofti* possessed a lifecycle involving the mosquito (see Chapter 3). Between 1889 and 1893, Theobald Smith (1859–1934) and Fred Kilborne (1858–1936) demonstrated that Texas cattle fever was transmitted by ticks from one animal to another, while in 1896 David Bruce (see Chapter 9) demonstrated the role of the tsetse fly in transmitting nagana from one animal to another.[5]

It remains unclear how many of these facts were known to Manson when he formulated his classical Goulstonian Lectures in 1896 (see Chapter 3) and, although he clearly stated that the 'necessary experiments cannot for obvious reasons be carried out in England', commended the hypothesis to the attention of medical men in India and elsewhere where 'malarial patients and suctorial insects abound'.

RONALD ROSS (1857–1932)

Ronald Ross (Figure 5.1) was born at Amora, in the Kumaön Hills at the foot of the Himalayas in the north-western provinces of India, on 13 May 1857 at the outbreak of the Indian mutiny. His father was General Sir Campbell Claye Grant Ross, KCB (1824–1892), of the Bengal Staff Corps of the Indian Army (whose

Ronald Ross
1908

FIGURE 5.1(a)   Ronald Ross (1857–1932), discoverer of mosquito involvement in human malaria and the complete life-cycle of avian malaria: (a) at the time of his Nobel Prize-winning work; (b) as an old man (reproduced courtesy of The Wellcome Library, London).

family had long connections with India), while his mother, Matilda Charlotte, was the eldest daughter of Edward Merrick Elderton, a London lawyer.

In 1865, Ross (the eldest of ten children) was sent to England for his education – initially at Ryde on the Isle of Wight, and later (in 1869) at a private boarding school in Springhill, near Southampton. He was at this time apparently deeply interested in zoology. Other early interests included geometry and the mathematics of music – interests which always remained with him. He received his medical education at St Bartholomew's Hospital in London, obtaining the MRCS in 1879, but unfortunately failing the Licentiate of the Society

FIGURE 5.1(b)

of Apothecaries of London (LSA). This left him unable to practise in Britain, but he was immediately employable as a ship's surgeon with the Anchor Line. However, in 1881 he did pass the LSA and, after taking a course at Netley, entered the Indian Medical Services (IMS) on 2 April of that year; it has been suggested that he was 'encouraged' in this direction by his father rather than by any inherent interest in becoming a 'zealous doctor'.

For several years he held temporary appointments with various Madras (now Chennai) regiments, or took medical duties at station hospitals in Madras, Burma (now Myanmar) and the Andaman Islands. During this time Ross studied the classics and the world's poets; he learnt Italian, French and German, and wrote poems, dramas, and novels (including *The Spirit of Storm* and *The Child of Ocean*). He also became even more deeply interested in mathematics.

In 1888 Ross left Madras for leave in England, and in 1889 he married Rosa Bessie, daughter of Alfred Bradley Bloxam, by whom he had two sons and two daughters (his wife, as well as a son and daughter, predeceased him). In 1888 he gained the diploma in public health (DPH), and studied bacteriology under Edward Emanuel Klein. His seminal work on malaria transmission (see below)[6] was carried out over a short period, between 1895 and 1898.

Later life

On 22 February 1899 Ross sailed from Calcutta (now Kolkata), arriving in London on 20 March, and became a lecturer (and later professor, 1902–1912) at the Liverpool School of Tropical Medicine at an initial salary of £250 per annum. He retired from the IMS on 31 July 1899.[7] As a result of his seminal discovery, Ross received numerous honours and was elected a Fellow of the Royal Society (FRS) in 1901 (being awarded its Royal Medal in 1909); in 1902 he received the Nobel Prize for Physiology or Medicine. He was appointed CB (1902), KCB (1911) and KCMG (1918).[8]

In 1912, Ross was appointed Physician in Tropical Diseases at King's College Hospital, London. At this time he turned his mind to eradication of the mosquito vector of malaria (he apparently 'often assured his friends' that his underlying motivation was basically humanitarian, and not scientific, with only partial success). He made visits to various countries, including Sierra Leone (1899–1900), West Africa (1901–2), Mauritius (1907–8), and Spain, Cyprus and Greece (1912). In the 1914–18 war Ross served in Alexandria and Salonika, and on 15 February 1917 he was appointed to the War Office as Consultant in Malaria. He was promoted to Colonel (Royal Army Medical Corps) in 1918.

For a short time, he established himself as a consultant in tropical diseases at 18 Cavendish Square, London W1. He was the second President (following Manson) of the (Royal) Society of Tropical Medicine and Hygiene (1909–11) (see Chapter 3). In 1926, the Ross Institute & Hospital for Tropical Diseases was founded at Putney Heath near London in his honour, with the object(s) of 'making it a centre of treatment and research …' (see Figure 5.2). It was opened by the Prince of Wales on 25 July 1926, and Ross became the first Director-in-Chief. He died there following a long illness, which had begun with a stroke, on 16 September 1932, and was buried at Putney Vale cemetery[9] (see Figures 5.3 and 5.4).

Every year from 1898 onwards, Ross commemorated 20 August (the date of his initial discovery in 1897) as 'Mosquito Day'. This event was last celebrated at the Ross Institute & Hospital for Tropical Diseases with an Indian lunch in 1931.

The day after Ross's death, *The Times* devoted its first leading article to his achievement(s):[10]

> though it is now many years since he made the discovery which is rendering the tropics safe for the white man, the importance of that discovery transcends any other since LISTER introduced his system of antisepsis. To say this, is not to belittle the pioneer work of

FIGURE 5.2 (a) Ross as Director of the Ross Institute & Hospital for Tropical Diseases, Putney; (b) an exterior view of the Institution (reproduced courtesy of the Ross Archive, London School of Hygiene and Tropical Medicine).

FIGURE 5.3   Interior of Holy Trinity Church, Putney West Hill, where Ross's funeral service took place.

FIGURE 5.4   Ross's grave at Putney Vale Cemetery.

MANSON, the 'father of tropical medicine'. MANSON broke new ground and offered suggestions to ROSS and others which were fruitful of good. But it was ROSS alone who achieved the discovery that the parasite of malaria fever is carried by the female anopheles mosquito. That discovery stands by itself ... One thing is certain – namely, that Ross's service glows with an imperishable lustre. He slew the dragon and delivered mankind from immemorial bondage. His name will live as long as the names of HARVEY and JOHN HUNTER, of JENNER and PASTEUR and LISTER, of MANSON and BRUCE and LEISHMAN.

ROSS'S WORK ON MALARIA

It was whilst Ross was on his second leave (from the IMS) in England in 1894 that his malaria work was conceived, although evidence exists that he had previously taken an interest in this disease – perhaps since 1889.[11] The prevailing theory had attributed malaria to 'a kind of miasma from marshes'; Manson considered that 'new human victims must acquire the germs by swallowing ... infected mosquitoes [which were, he considered, in some way implicated in the disease] or germs passed by them into water'. He met Manson, having been introduced to him by Professor Kanthack (a pathologist at St Bartholomew's Hospital), at his home, 21 Queen Anne Street, London, on 9 April of that year. At that time, Ross was quite unable to visualize Laveran's organism. When Ross first called at Manson's house, 'the father of tropical medicine' was away; however, he wrote promptly (the same day in fact) indicating that Ross's difficulty in visualizing malaria parasites had resulted from a technical inadequacy.[12]

Ross, by this time a trained bacteriologist and accomplished microscopist, returned to Secunderabad, India, on 24 April 1895. It was two years and four months later that his seminal discovery was made. In early 1896, Manson had expanded his 'mosquito theory' into three classical Goulstonian Lectures which were delivered to the Royal College of Physicians (see Chapter 3).[13]

On 20 August 1897 in Secunderabad, Major Ronald Ross (1857–1932) (see Figures 5.5–5.7) visualized pigmented 'coccidia' (oocysts) of Laveran's malarial parasite in the stomach of an uncommon type of female mosquito – the 'dapple-winged' variety, now designated *Anopheles* spp. – which had ingested blood containing 'crescents' (gametocytes) from a patient suffering from malaria. That evening he wrote to his wife: 'I have seen something very promising indeed, in my new mosquitoes,' and he apparently scribbled the following unfinished verses in one of his *In Exile* notebooks in pencil:

This day designing God
    Hath put into my hand
A wondrous thing. And God
    Be praised. At His command,
I have found thy secret deeds
    Oh million-murdering Death.

I know that this little thing
    A million men will save –

FIGURE 5.5   The laboratory in Secunderabad in which Ross worked (reproduced courtesy of the Ross Archive, London School of Hygiene and Tropical Medicine).

> Oh death, where is thy sting?
> Thy victory oh grave?

Ross recorded in his *Memoirs*:[14]

Then, or a few days later, I wrote the following amended verses on a separate slip of paper:

This day relenting God
Hath placed within my hand
A wondrous thing; and God
Be praised. At His command,

Seeking His secret deeds
With tears and toiling breath
I find thy cunning seeds,
O million-murdering Death.

I know this little thing
A myriad men will save,
O Death, where is thy sting?
Thy victory, O Grave?

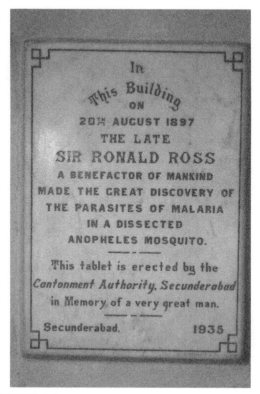

FIGURE 5.6 Plaque on the exterior wall of Ross's laboratory.

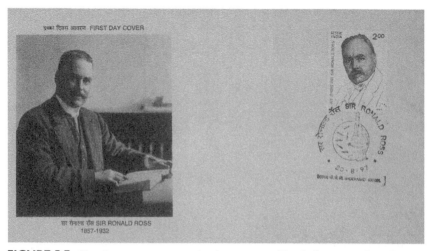

FIGURE 5.7 First-day cover commemorating the centenary of Ross's discovery at Secunderabad.

LABORATORY AT CALCUTTA. SURGEON-MAJOR ROSS, MRS. ROSS, MAHOMED BUX AND LABORATORY ASSISTANTS, 1898.

FIGURE 5.8 The Calcutta laboratory in which Ross elucidated the life-cycle of avian malaria (*Proteosoma* sp.) (reproduced courtesy of The Wellcome Library, London).

Ross subsequently published an account of these 'cysts' in the *British Medical Journal* for 18 December 1897.[15]

This work was brought to a sudden halt when Ross was posted on 26 September 1897 from Secunderabad to Kherwara in Rajputana, where there was no human malaria.[16] Only the intervention of Manson, in London, with the Secretary of State for India allowed Ross to resume his work. In the following year, Ross embarked (at Manson's suggestion) on a study of *Proteosoma* sp., an avian parasite, at Calcutta[17] (see Figure 5.8). On 20 March 1898 he observed similar pigmented cysts and, later, subsequent stages ('germinal rods' or sporozoites) in the salivary glands of a mosquito (presumably *Culex fatigans*) which had fed on an infected lark.[18] The full cycle of development of

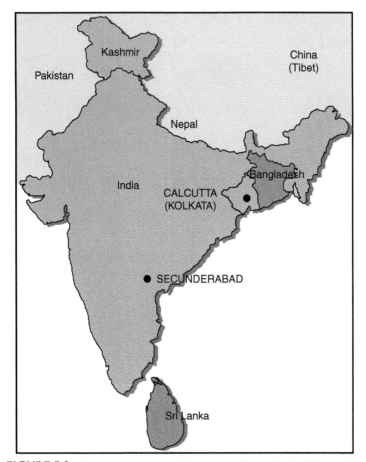

FIGURE 5.9   Map showing the sites of Ross's major discoveries in 1857 and 1858.

the avian parasite became clear on 9 July, and on 28 July 1898 Manson (who had received telegraphed news) announced the result of Ross's experiment(s) – carried out in Calcutta – at the Edinburgh meeting of the British Medical Association.[19]

Figure 5.9 shows the sites of Ross's major discoveries at Secunderabad and Calcutta, respectively. Ross subsequently claimed that 'few people in England knew anything about [his] Indian work, because it had been cleverly pirated by a Roman savant near the end of 1898'. When he 'received the Nobel prize for physiology or medicine in 1902 [he wrote], I did not feel called upon to divide it with Manson even if I had thought it proper to do so at all'.[20] This somewhat paranoid response it seems was a typical reaction of the great polymath – with only a passing interest in medicine.[21]

Manson's role in Ross's discovery

It is difficult to assess the relative part played by Manson in Ross's discoveries. There is, however, no doubt that these monumental events immediately exploded the view, then universally held, that human malaria was contracted from contaminated air and/or water. Although Manson did not 'invent the mosquito theory', by extrapolation from his work on lymphatic filariasis carried out in China, he undoubtedly gave it 'shape and life'. There can, however, be no doubt of Manson's pivotal role in providing the scientific background and intellectual impetus. It is of interest that only the year *after* Ross received the Nobel Prize was Manson awarded a knighthood. Manson also acted as Ross's mentor and, by virtue of his by then elevated position in the British establishment, was influential in obtaining a posting (in 1898) for Ross to Calcutta.

Following his 1898 elucidation of the life-cycle of avian malaria, Ross was 'ordered away from Calcutta to investigate kala-azar, the nature of which was then unknown'(see Chapter 12). This left him unable to confirm his avian findings in human malaria.[22] The life-cycle of *Plasmodium* spp. which infect humans was in fact initially worked out (in December 1898) by Battista Grassi (1854–1925) and his Italian colleagues, and this led to a bitter controversy, which lasted 20–30 years, between Ross and Grassi on the question of priority; it probably harmed them both and 'served no useful purpose'.[23] Mosquito transmission of the human parasite was finally confirmed by Manson's colleagues working in the Roman Campagna; this work was published in 1900.[24]

In his *Memories of Sir Patrick Manson* (1930), Ross described briefly his work at Secunderabad and Calcutta. He wrote, 'In July and August 1898, I found something which had never been suspected by Manson or by myself', and he then described the mode of mosquito infection of birds, i.e. via the 'salivary or poison-glands'. Although Ross was apparently grateful to Manson for his appointment to the Liverpool School, he admits he did not like teaching and quotes George Bernard Shaw ('He who can, does. He who cannot, teaches') and also Sir David Bruce ('They who require to be taught are seldom worth teaching').[25]

THE ITALIAN MALARIOLOGISTS

Several Italians, most of them based in Rome, have already been brought to the reader's attention in Chapter 4. Bignami, in a two-part article published in the *Lancet* following Manson's Goulstonian Lectures (see Chapter 3), argued against the idea of infection through drinking water and, like many others, suggested that the bite of the mosquito might be responsible. He cited Celli (see Chapter 4) as having demonstrated experimentally that water could not be the means of infection; he also criticized the soil and wind theories. He had, he wrote, demonstrated with Bastianelli in 1893–4 that extremely small quantities of blood from a malarious patient, when inoculated (by the intravenous or subcutaneous route) into

FIGURE 5.10   Giovanni Battista Grassi (1854–1925), participant in the claim for priority in the discovery of the life-cycle of human malaria (reproduced courtesy of The Wellcome Library, London).

another individual, produced the disease.[26] Bignami did not, however, allude to Gerhardt's demonstration, in 1884 (i.e. a decade earlier), that blood from a malarious subject, when injected into a healthy individual, produces malaria.[27]

On 22 September 1898, Battista Grassi (1854–1925; Figure 5.10) communicated his finding that malaria was transmitted from individual to individual *only* by the bite of *Anopheles* and not *Culex*. This was the first demonstration of the complete malaria cycle in *Homo sapiens*.[28]

Poser and Bruyn, following a careful analysis of the evidence, have concluded that:[29]

> Grassi became the first to describe the complete life cycle of *Plasmodium falciparum* while … Bastianelli and … Bignami did the same for *P. vivax* and later for *P. malariae*.

The priority controversy

There is absolutely no question that Ross, by resorting to avian malaria (which had previously been studied by McCallum the same year),[30] discovered the full life-cycle of avian malaria (*Proteosoma* spp.) by *Culex fatigans* in Calcutta in 1898. This was confirmed by C W Daniels (see Chapter 9) on behalf of the Royal Society.[31] The Italian malariologists Grassi, Bignami and Bastianelli, however, elucidated the complete cycle of human malaria (*Plasmodium* spp.) the following year (see above).

The controversy centres upon the extent to which Ross's discovery influenced the Italian workers, most notably Grassi. A major problem seems to have stemmed from Ross's somewhat paranoid personality; Chernin has written that he was 'quick to take offence and [was] capable of magnifying a petty affair out of all proportions', while others have considered him either 'chronically maladjusted', or 'a tortured man'.[32] He also turned against Manson (who had pointed him to his seminal discoveries) in his latter years. A financial ambition was another quirk; he compared his 'discovery' to that of Edward Jenner, who received £30000 from the British Parliament for 'discovering' smallpox vaccination.[33]

Ross was especially hostile to Grassi, about whom he later wrote:

> Grassi discovered nothing in connection with malaria. The [only] Italian results of value regarding the mosquito-malaria theory were due to Bignami, who as he himself had said, used Grassi as an entomological assistant to identify his mosquitoes.

Grassi in turn described Ross's Nobel Prize-winning observations as 'inconclusive, dubious and positively misleading'. As Poser and Bruyn have written:[34]

> He also said that he, Grassi, with Bignami, deserved the credit for showing that malaria was conveyed to man by mosquito bites, and finally he claimed emphatically that elucidating the mode of transmission of malaria had been accomplished in Italy as a result of research begun by himself, continued with the assistance of Bignami and Bastianelli. The credit for the final completion, however, belonged to Grassi alone. Ross remained embittered for the rest of his life as a result of Grassi's fame.

Furthermore, Ross wrote, in September 1900:[35]

> Looking at all these facts I find it impossible to admit that the work of the Italians possesses the fundamental originality which they endeavour to claim for [it]. On the contrary, it was almost entirely stimulated, sustained and guided by the work of Manson, MacCallum and myself, especially by the general solution of the problem contained in my investigation of the malaria of birds. Had it not been for the original work, I believe that the Italians would have been just as ignorant regarding the mosquito theory today as they were ten years ago. On the other hand had they never touched the subject, I am confident the work would have been just as thoroughly completed by Koch, the [Malaria] Commission of the Royal Society, myself and others. In a word, the whole of the Italian work depended on my discovery of the zygotes. I repeat, as I have said before and as I know to be the case, that when they took up the work it no longer presented any serious difficulties. Besides being unable to personally admit their claims, I cannot help thinking that they have pushed their efforts in support of these claims too far – that they have endeavoured to obtain by literary artifices and polemical dexterities a credit to which their researches do not entitle them.

Poser and Bruyn continue, in an excellent account of the controversy:

> Grassi's statement that my [Ross's] researches on the malaria of birds were published almost contemporaneously with his first note is both in meaning and in substance absolutely untrue ... Stripping off the plagiarisms and pretensions which Grassi had so carefully plastered upon his work to increase its bulk in the eyes of the world, we shall find on careful scrutiny that but little remains. I cannot observe a single idea which is at once sound, original and fundamentally important, contained in the contributions of Grassi *solus*. Grassi's writings may be justly said to consist of his series of grave errors, another series of rash hypotheses, and some sound efforts founded on another's labors [sic]. Whatever his achievements in other branches of science may be, his claim in regard to malaria may be defined as being those of an energetic, dexterous, and unscrupulous writer who has discovered the discovery of another man.

Edmonston Charles (1834–1906), a staunch supporter of Ross, wrote to him in November 1898:[36]

> when he [Charles] called on Grassi in his laboratory, the latter had before him a copy of *The British Medical Journal* containing Ross's December 18, 1897, paper on pigmented cells in dappled-winged mosquitoes, and declared that he had no doubt from Ross's description that the mosquitoes were *Anopheles claviger*.

But that was not all. Ross cited a letter from C W Daniels dated October 1900 which stated that when Daniels arrived in Calcutta in December 1898 (see above), Ross had shown him specimens of the grey, brindled and dapple-winged mosquitoes; the latter were the *Anopheles* mosquitoes. Ross also showed him a specimen of the stomach of a mosquito with what are now known as zygotes. This was the stomach of a dapple-winged mosquito similar to 'those he had shown me which had been fed on a patient with crescents in 1889'.[37]

Meanwhile, Koch and Laveran both sided with Ross, and Lister, speaking at a British Medical Association meeting in Glasgow, also voiced his support:[38]

> [Lister, in a report of the meeting] wished to bear testimony to the qualities which had enabled Major Ross to bring about this great discovery, because the discovery of the development of the parasites in the mosquito was due solely and simply to Major Ross, who had shown admirable scientific acumen and zeal and perseverance. At the same time he had – very differently indeed from some Italian investigators – shown absolute candor, perfect openness of mind, and a readiness to recognize the work of others.

As Poser and Bruyn have, however, also pointed out, almost a year before, in December 1900, Lister had praised Grassi's contributions and presented the Darwin Medal to him!

In summing up the controversy regarding priority, George Nuttall (1862–1937) wrote:[39]

> When Grassi, Bignami and Bastianelli in Rome, stimulated and directly guided by the epoch-making discoveries of Ronald Ross on malaria, confirmed and extended the latter's work, the antagonistic attitude adopted by Koch [see above] was extreme. It was generally held in Berlin that the work of the Italian observers was worthless and it was not until I had demonstrated some of Grassi's specimens in Berlin after having gone to Rome to

fetch them, that a measure of credence was given by some authorities in Germany to the reported results of the Italian researchers.

However, in a defence of Grassi, Nuttall wrote:

It must be assumed that Koch had not seen [at the time of his assessment] any of the preparations made by Grassi and his colleagues, as late as February 10, 1901.

'It is clear', he concluded:

that full credit must be given to Grassi and his colleagues for having been the first to demonstrate that, for the human malaria parasite, the cycle of development in *Anopheles* corresponds exactly to that demonstrated by Ross for *Proteosoma* and *Culex*. On the other hand, there can be no question but that Ross observed the early stages of the development of one species of human malaria (*praecox*) in *Anopheles*.

Manson-Bahr too came to Grassi's defence, but he was of course not a great supporter of Ross anyway (see below), and had previously written critically of him. He concluded by writing:[40]

Surely a bare recital of all this places the work of these Italians on a plane only second to that of Ronald Ross. [Manson-Bahr] also added an appreciative note regarding ... Nuttall's role. He did not make any direct scientific contribution to malaria but, by his accuracy, wide knowledge and technical ability played a very considerable part in the malaria story, and knowing all parties, he may be said to have acted as a referee and held the scales of priority.

In 1978, Gordon Harrison admirably summarized the controversy:[41]

It makes no sense of justice to couple the names of Ross and Grassi together as co-discoverers ... without noting the very large difference between the explorer at the helm and those who rode his decks and helped make a landing. Both men were great achievers, but that one became a Nobel laureate and the other a Roman senator satisfied neither; perhaps Ross and Grassi deserved each other.

IS THE MOSQUITO AFTER ALL RESPONSIBLE FOR TRANSMITTING MALARIA?

Following the Ross/Grassi discoveries, the fact that malaria occurred in areas in which there were few or no *Anopheline* mosquitoes, and *vice versa*, continued to perplex entomologists. In 1929, S P James (1870–1946) wrote that in many areas in England in which malaria was then absent, there were *more Anopheline* mosquitoes than in endemic ones.[42] However, L W Hackett (1884–1962) pointed out in 1937 that in Europe the change from cereal to animal husbandry had been associated with a diminution in the prevalence of the malaria-carrying *Anopheles* mosquitoes. There were, he concluded, two races within *A. maculipennis*: one fed on domestic animals and the other on man. This idea became widely accepted, but many unanswered questions remain.[43]

SIR PHILIP MANSON-BAHR'S ASSESSMENT OF SIR RONALD ROSS

Dr P E C Manson-Bahr (1911–96),[44] son of Sir Philip (1881–1966), sent a letter to me dated 18 February 1989 enclosing a photocopy of some notes (marked *private*) concerning Ross, which were written by his father in 'about 1960':

I first was introduced to Ronald Ross in Liverpool in 1899 when I was 18 years of age and was at that time assistant in my grandfather's office. Ross was then a social lion and was invited out to numberless dinner parties in these circles in which I moved. The large hospitals in Liverpool – the Great Northern and Stanley [?] – elected him to their staffs on his peak reputation, but he soon resigned as he cared little for clinical medicine.

Some two years later, just before I went up to Cambridge, there was an exciting event in Liverpool. I can still hear the ragged, mud-bespattered newspaper boys shouting '*Echo Echo*' (evening newspaper). There was a scandal about two doctors – David Bruce & [see Chapter 9] & Ronald Ross at the Bee Hotel in Abergele. They appeared to have enjoyed a good dinner and had slipped in next door for a haircut. During the operation the barber inadvertently snipped off a portion of Bruce's ear. Bruce, who had been a good boxer, retaliated & knocked the unfortunate barber out. Liverpool was all agog & they were both fined & the magistrate had some little homilies to say.

As a Professor in Liverpool Ross did not appear to be a great success for he was not a facile lecturer & all the original teachers at the School who were there when he came from India – Warrington Yorke, J. W. W. Stephens & Newstead all were of the same opinion. Ross knew little about haematology, & was not familiar with the morphology of malaria parasites & unable to dissect mosquitoes. How to satisfactorily explain all this is by no means easy, after the very great original work he had accomplished & that won him the Nobel Prize. De Cuif, in the 'Microbe Hunters' wrote that his Indian assistant Mohammed Bux was the man who performed all the dissections for him in Calcutta & was indeed the main factor in the situation. Of all this I cannot pretend to know anything. All that I do know is when I heard him give a [?] lecture. I had quite a shock & disappointment when he held forth at the Medical School at the Albert Dock. He mumbled and stuttered to such an extent that I failed to understand anything he said. At this time, he had achieved some notoriety by enclosing two trypanosome cases from West Africa in a cold room at the Royal Infirmary in Liverpool with the temperature at freezing point so as to depress the fever, with the result that both had died of pneumonia.

It has always been a puzzle to explain the relationship of Ronald Ross to Sir Patrick Manson. In 1894 when Ross came to London to sit at Manson's feet, Manson took Ross by the arm as they walked in Oxford Street saying 'Do you know Ross, I believe that mosquitoes carry malaria'. As Ross wrote at the close of his Nobel address in Stockholm in 1901 [*sic*] 'Had it not been for Manson he should still be looking for the cause of malaria in water or in air'. After his appointment in Liverpool as Professor at Manson's instigation, he continued to complain publicly that his reward was inconsistent with his great services. Then he resigned. He began to write letters to *The Times* and gradually became more hostile to Manson. Things became so strained that in 1909 he & Bruce apparently by design, became so critical at a meeting of the Tropical Diseases Committee of the Royal Society that Sir Patrick took their remarks as an affront to the British government & as its representative he left the room and never returned in spite of the remonstrances & assurances of the President – Sir Archibald Geikie [1835–1924[45]].

Later in 1911 Sir Patrick had a bad attack of gout. Ross had just published a letter belittling the part Sir Patrick had played in the malaria story. He then said to me 'Go upstairs & bring down a brown paper parcel & wrap it up well & address it to Sir Ronald Ross'. This packet contains the whole correspondence between us on the malaria problem. I wish him to have them as 'I never wish to see him or speak to him again'. Eventually Ross sold the letters to Lady Houston for £7,000.

Then, on Sir Patrick's retirement, Ross came to London. He had bought Sir Stephen MacKenzie's [1844–1909[46]] house in Cavendish Square & put up his plate. His practice was not successful. It was rumoured that his quinine mixtures were actually explosive!! At any rate he found himself soon in financial difficulties.

Early in 1913 Sir Patrick was staying with me in Ceylon [now Sri Lanka] and one day a cable arrived from a solicitor in Liverpool to the effect that Sir Patrick had libelled Sir R Ross & he demanded an apology be made on the front page of the *Times*. As Sir Patrick was away from Nouvera Eliza [?] & with me in the jungle Lady Manson opened the telegram. She was dumbfounded. When Sir Patrick returned to Nouvera Eliza [?] he was in great distress & he had no idea what it was all about. It then appeared that when Ross retired from his Professorship in Liverpool a substitute had to be appointed. The name of Dr W Prout [?–1939] had been put forward & he had written to Sir Patrick for a testimonial. The letter which I still have was written in Lady Manson's hand to the effect that Prout, who had been PMO Gambia and had a large experience in W Africa was to be recommended as a clinician as it was most important that Liverpool should have the services of a good physician. By some obtuse process of thinking Ross took this as a slight on himself & somehow or other got some others to back him up – notably one called [Robert] Fielding-Ould [1872–1951]. However in the end Sir Patrick wrote to the solicitor & enclosed £5-5 to pay his fee with his apology for causing any trouble quite unwittingly. The solicitor was so decent as to return the fee with his own apologies for having been let in to such an unwarrantable charge.

During the Great War [1914–18] I came across Sir Ronald Ross at various times. He was the authority who decided that the dysentery in Gallipoli was amoebic in origin, but there can be little doubt that he mistook the refractile macrophage cells of the bacillary exudate as *E. histolytira*. He bore a great responsibility for what followed.

Then again in Nov. 1915 he was sent to Salonika to make a preliminary survey of malaria and the area to be occupied by the Expeditionary Force. He was accompanied by F W O'Connor.[47] As the malaria season was at an end he did not see any malaria but few anopheles. He never reached the Struma river. He reported to the War Office that malaria in Salonika was not a menace. The result all the world now knows!!

I did not see him again till June 1919 when I returned from Egypt bearing Fairley's [see Chapter 15] papers on the schistosomiasis antigen for publication in the Jl. RAMC of which he was then the Editor. He was not very polite and agreeable & handed me back the script of both papers which he described as Bosh!

ROSS'S LATER THOUGHTS ON THE FUTURE OF TROPICAL MEDICINE

The underlying theme of Ross's Presidential Address to the (Royal) Society of Tropical Medicine and Hygiene in 1908 was: our duty to ascertain that the fruits of our researches are implemented. Following his seminal discovery, therefore, the major thrust of his future work was centred on prevention – with regard to not only malaria, but other tropical diseases also. Ross suggested therefore that there should be separate courses devoted to tropical sanitation. Cure is all very well, he claimed, but prevention is more effective. Examples of *curative* medicine were atoxyl in trypanosomiasis, etc., quinine in malaria, and serum therapy in plague.

Regarding sanitation in the tropics, important priorities were:

1. The housing of the poor. Ross spoke of 'bamboo-shelters and mud-huts swarming with vermin, noisome yards strewn with rubbish, unspeakable latrines, broken gutters and poisonous wells'.
2. Removal of refuse
3. Attention to the water supply.

All large villages and rural areas, he claimed, should have 'a qualified sanitary inspector', and 'every tropical country or colony must have a whole-time chief sanitary officer'.

Ross, in fact, advocated whole teams for a large territory, paying attention to:[48]

1. Water and food supplies
2. Drainage
3. Conservancy/house sanitation
4. Statistics
5. Mosquito-borne diseases
6. Epidemics.

NOTES

1 C Lancisi. *De noxiis paludum efflaviis*. Romae, 1717: Salvioni; B Rush. *An Enquiry into the Various Sources of Summer and Autumnal Disease*. Philadelphia, 1805: J Conrad; L Beauperthuy. *La fiebre amarilla*. Gaceto Official Cumana 1854, 23 May; J Jaramillo-Arango. The history of the mosquito-malaria theory. *Med Bookman Hist* 1948, 2: 345–8, 391–5; C M Poser, G M Bruyn. *An Illustrated History of Malaria*. London, 1999: Parthenon Publishing Group, p. 37.

2 *Op cit.* See note 1 above (Poser, Bruyn), pp. 39–40.

3 A F A King. Insects and disease – mosquitoes and malaria. *Popul Sci Monthly* 1883, 23: 644–58.

4 C Flügge. *Grundriss der Hygiene*. Leipzig, 1891: Viet; R Pfeiffer. *Coccidienkrankheit der Kaninchen*. Berlin, 1892: Hirschwald; *Op cit.*, see note 1 above (Poser, Bruyn), p. 39.

5 P Abildgaard. *Pferde-und Vieharzt in einem kleinen Auszuge*. Wien, 1785: von Trattern; A Fedchenko. On the structure and reproduction of *Filaria medinensis*. *Am J Med* 1971, 20: 511–23.

6 R Ross. *Memoirs, with a Full Account of the Great Malaria Problem and its Solution*. London, 1923: John Murray, p. 547; Anonymous (obituary). Sir Ronald Ross: the conquest of malaria. *Times, Lond* 1932, 17 September; Anonymous (obituary). Sir Ronald Ross KCB, KCMG, MD, LID, DSc, FRS. *Br Med J* 1932, ii: 609–11; Anonymous. Ronald Ross 1857–1932. His life and work. *Lancet* 1932, ii: 695–7; G H F Nuttall. Sir Ronald Ross (1857–1932). In: *Obituary Notices of Fellows of the Royal Society*. London, 1933: Harrison and Sons, pp. 108–15; C M Wenyon. Obituary. Colonel Sir Ronald Ross, KCB, KCMG MD DSc, LID, FRCS, FRS, IMS (retd) 1857–1832. *Trans R Soc Trop Med* 1933, 26: 473–8; H H Scott. Ronald Ross (1857–1932). In: *A History of Tropical Medicine*, Vol. 2. London, 1939: Edward Arnold, pp. 1086–90; G C Cook. Ronald Ross (1857–1932): 100 years since the demonstration of mosquito transmission of *Plasmodium* spp. – on 20 August 1897. *Trans R Soc Trop Med Hyg* 1997, 91: 487–8. W F Bynum. Ross, Sir Ronald (1857–1932). In: H C G Matthew, B Harrison (eds), *Oxford Dictionary of National Biography*, Vol. 47. Oxford, 2004: Oxford University Press, pp. 842–6; W F Bynum, C Overy (eds). *The Beast in the Mosquito: the correspondence of Ronald Ross and Patrick Manson*. Amsterdam, 1998: Rodopi B, p. 528; E R Nye, M E Gibson. *Ronald Ross, Malariologist and Polymath: a biography*. London, 1997: MacMillan Press Ltd, p. 316;

R Ross. *The Great Malaria Problem and its Solution: from the memoirs of Ronald Ross*. London, 1988: Keynes Press, p. 236; R L Mégroz. *Ronald Ross: discoverer and creator*. London, 1931: George Allen & Unwin Ltd, p. 282; J Kamm. *Malaria Ross*. London, 1963: Methuen & Co Ltd, p. 181; E F Dodd. *The Story of Sir Ronald Ross and his Fight against Malaria*. London, 1956: MacMillan & Co Ltd, p. 81; J O Dobson. *Ronald Ross Dragon Slayer: a short account of a great discovery and of the man who made it*. London, 1934: Student Christian Movement Press, p. 162; J Rowland. *The Mosquito Man: the story of Ronald Ross*. London, 1958: Lutterworth Press, p. 150; G C Cook. Ronald Ross (1857–1932): 100 years since the demonstration of mosquito transmission of *Plasmodium* spp. – on 20 August 1897. *Trans R Soc Trop Med Hyg* 1997, 91: 487–8. See also N Hawkes. DNA scientists join war against malaria. *Times, Lond* 1997, 13 August: 7.

7   *Op cit*. See note 6 above (Anonymous, *Lancet* 1932).

8   *Op cit*. See note 6 above (Nuttall).

9   *Op cit*. See note 6 above. See also: G C Cook. Aldo Castellani FRCP (1877–1971) and the founding of the Ross Institute & Hospital for Tropical Diseases at Putney. *J Med Biog* 2000, 8: 198–205; G C Cook. A difficult metamorphosis: the incorporation of the Ross Institute & Hospital for Tropical Disease into the London School of Hygiene and Tropical Medicine. *Med Hist* 2001, 45: 483–506; G C Cook. The Albert Dock years: 1899–1920. In: *From the Greenwich Hulks to Old St Pancras: a history of Tropical Disease in London*. London, 1992: Athlone Press, pp. 163–217; G C Cook. The grave of Sir Ronald Ross FRS (1857–1932). *Lancet* 1999, 354: 1128.

10 Anonymous [leading article]. Sir Ronald Ross. *Times, Lond* 1932, 17 September.

11 G C Cook. Manson's demonstration of the malaria parasite 100 years ago: the major stimulus for Ross's discovery? *J Infect* 1994, 28: 333–4.

12 *Ibid*. See also P H Manson-Bahr, A Alcock. *The Life and Work of Sir Patrick Manson*. London, 1927: Cassell, p. 273.

13 P Manson. The Goulstonian Lectures on the life-history of the malaria germ outside the human body. *Lancet* 1896, i: 695–8, 751–4, 831–3; Anonymous. Reports of Societies: Royal Medical and Chirurgical Society. *Br Med J* 1896, i: 530–31. See also G C Low. A retrospect of tropical medicine from 1894 to 1914. *Trans R Soc Trop Med Hyg* 1929, 23: 215–17.

14 R S Desowitz. *The Malaria Capers: more tales of parasites and people, research and reality*. London, 1991: W W Norton & Co, p. 288; *Op cit*. See note 6 above (Ross), p. 226.

15 *Op cit*. See note 1 above. See also Anonymous (editorial). Ronald Ross. *Br Med J* 1932, ii: 597; G C Cook. The great malaria problem. *Natl Med J India* 1989, 2: 57.

16 R Ross. On some peculiar pigmented cells found in two mosquitoes fed on malarial blood. *Br Med J* 1897, ii: 1786–8.

17 *Op cit*. See note 6 above (Nuttall, Wenyon).

18 R Ross. Report on the cultivation of Proteosoma, Labbé, in grey mosquitoes. *Indian Med Gaz* 1898, 33: 401–8, 448–51.

19 P Manson. The mosquito and the malaria parasite. *Br Med J* 1898, ii: 849–53. See also: *Op cit*. See note 6 above (Nuttall); G C Cook. Mosquito involvement in the malaria life-cycle. *J Med Biog* 1998, 6: 182–3.

20 R Ross. *Memories of Sir Patrick Manson*. London, 1930: R Ross, p. 26.

21 E Chernin. Sir Ronald Ross vs Sir Patrick Manson: a matter of libel. *J Med Hist* 1988, 43: 262–74; E Chernin. Sir Ronald Ross, malaria, and the rewards of research. *Med Hist* 1988, 32: 119–41.

22 *Op cit*. See note 6 above (Wenyon, Christophers); note 15 above (Anonymous); *Op cit*. See note 1 above (Poser, Bruyn), pp. 41–2.

23 *Op cit*. See note 6 above (Nuttall, Wenyon, Christophers). See also W C Campbell. Ronald Ross and Battista Grassi. *J Med Biog* 1998, 6: 182.

24 P Manson. Experimental proof of the mosquito-malaria theory. *Br Med J* 1900, ii: 949–51; G C Cook. *From the Greenwich Hulks to Old St Pancras: a history of Tropical Disease in London*. London, 1992: Athlone Press, p. 338.

25 *Op cit*. See note 20 above.

26 A Bignami. Hypotheses as to the life-history of the malarial parasite outside the human body. *Lancet* 1896, ii: 1363–7, 1441–4.

27 C Gerhardt. Ueber Intermittens-Impfungen. *Zeitschr Klin Med* 1884, 7: 372–7.

28 B Grassi. Rapporti tra la malaria e peculiari insetti (Zanzaroni e zanzare palustri). *Policlinico* 1898, 5: 469–76; B Grassi. Mosquitoes and malaria. *Br Med J* 1899, ii: 748–9.

29 *Op cit.* See note 1 above (Poser, Bruyn), p. 45.

30 W G MacCallum. On the flagellated form of the malarial parasite. *Lancet* 1897, ii: 1240–41.

31 G C Cook. Charles Wilberforce Daniels FRCP (1862–1927): underrated pioneer of tropical medicine. *Acta Trop* 2002, 81: 237–50.

32 *Op cit.* See note 21 above. See also S Pampiglione, S Giannetto. The Grassi–Calandruccio controversy. Who is wrong? Who is right? *J Med Biog* 2001, 9: 81–6.

33 *Op cit.* See note 21 above. (*Med Hist*).

34 *Op cit.* See note 1 above (Poser, Bruyn) p. 48.

35 R Ross. Le scoperte del Prof. Grassi sulla malaria. *Policlinico* 1900, 29 December.

36 *Op cit.* See note 1 above (Poser, Bruyn) p. 48.

37 *Op cit.* See note 31 above.

38 Anonymous. *Times, Lond* 1901, 19 September.

39 C Nuttall. On the question of priority with regard to certain discoveries upon the aetiology of malarial diseases. *Q J Microscop Sci* 1901, 44: 429–41.

40 P Manson-Bahr. The story of malaria: the drama and actors. *Intern Rev Trop Med* 1963, ii: 329–90.

41 G Harrison. *Mosquitoes, Malaria and Man: a history of the hostilities since 1880*. New York, 1970: Dutton; *Op cit.* See note 1 above (Poser, Bruyn), p. 50.

42 S P James. The disappearance of malaria from England. *Proc R Soc Med* 1929, 23: 71–87.

43 L W Hackett. *Malaria in Europe*. London, 1937: Oxford University Press; L Bruce-Chwatt, J de Zulueta. *The Rise and Fall of Malaria in Europe*. Oxford, 1980: Oxford University Press, p. 240.

44 G C Cook. Manson-Bahr, Philip Edmond Clinton. *Munk's Roll*, Vol. 10. London: Royal College of Physicians, pp. 328–30.

45 D Oldroyd. Geikie, Sir Archibald (1835–1924). In: H C G Matthew, B Harrison (eds), *Oxford Dictionary of National Biography*, Vol. 21. Oxford, 2004: Oxford University Press, pp. 721–3.

46 H D Rolleston, H Series. MacKenzie, Sir Stephen (1844–1909). In: H C G Matthew, B Harrison (eds), *Oxford Dictionary of National Biography*, Vol. 35. Oxford: Oxford University Press 2004: 618; Anonymous. MacKenzie, Sir Stephen. *Munk's Roll*, Vol. 4. London: Royal College of Physicians, pp. 266–7.

47 Anonymous. O'Connor, Francis Wm. *Medical Directory*. London, 1920: J & A Churchill, p. 285; P Manson-Bahr. Dr Francis W O'Connor. In: *History of the School of Tropical Medicine in London (1899–1949)*. London, 1956: H K Lewis, pp. 188–9.

48 R Ross. The future of tropical medicine. *Trans Soc Med Hyg* 1909, 2: 272–88.

6

Carlos Finlay (1833–1915): yellow fever research in southern America

Implication of the mosquito in the transmission of lymphatic filariasis and malaria probably laid the foundation for the pioneering contributions to understanding of the cause of yellow fever. Immediately following this scientific proof was the influence exerted by the disease in slowing construction of the Panama Canal. It should be noted that it is often difficult or impossible to differentiate yellow fever from the severe viral hepatitides in historical accounts.

Yellow fever ('yellow jack') is a viral disease which was an enormous hazard facing the West African Squadron.[1] The *Flying Dutchman* legend, and the *Rime of the Ancient Mariner*[2] by S T Coleridge (1772–1834),[3] provide examples of this disease afflicting non-immune mariners. However, in the seventeenth and eighteenth centuries the disease was far more widespread than the areas of Africa and South America in which it is presently endemic. It was, for example, not unusual in North America; Benjamin Rush (1746–1813)[4] described an epidemic in Pennsylvania in 1793,[5] and epidemics have occurred in several European countries in the past. The infection was rampant in the West Indies, Jamaica included, from about 1850; however, exposed water, the natural breeding habitat of *Aëdes aegypti*, diminished as sanitation improved and drinking water was supplied in pipes.[6]

MODE OF INFECTION

As with malaria, miasmas were long felt to be the source of infection. In the Philadelphia outbreak, Rush believed that the epidemic was caused by damaged coffee beans lying on the wharves! Towards the end of the nineteenth century, turning over of soil in construction work was held responsible.

Joseph Nott (1804–73) was probably the first to suggest, in 1848, that mosquitoes were involved in transmitting the disease.[7] However, it was Carlos Finlay (1833–1915; Figure 6.1)[8] who indicated categorically that mosquitoes were involved in person-to-person transmission. Finlay was the son of a Scottish father and French mother who emigrated to Cuba when Carlos was an infant. After graduation at Philadelphia he took up practice in Havana, and in 1881 delivered a lecture on the role of the *A. aegypti* mosquito in transmission to the International Sanitary Commission at Washington; this was before news of

FIGURE 6.1   Carlos Finlay (1833–1915), who first incriminated *Aëdis aegypti* in the transmission of yellow fever (reproduced courtesy of The Wellcome Library, London).

Manson's 1877 work in Amoy on mosquito involvement in lymphatic filariasis transmission (see Chapter 3) had reached Cuba. Finlay backed up his hypothesis with a great deal of clinical observation which, although largely ignored today, virtually clinched the role of *A. aegypti* in transmission.[9]

The American Yellow Fever Commission

In 1898, American soldiers were suffering a high mortality rate from the disease in Cuba, and therefore in 1900 the US Government appointed a Commission to investigate the problem. This was headed by Walter Reed (1851–1902; Figure 6.2),[10] Professor of Bacteriology at the Army Medical School. James Carroll (1854–1907; Figure 6.3), Jesse Lazear (1866–1900; Figure 6.4) and Aristide Agramonte (1868–1931) were the other members.[11] They had been preceded in Cuba by a British Yellow Fever Commission, consisting of W Myers (1872–1901; Figure 6.5) and H E Durham (1864–1945) from the Liverpool School of Tropical Medicine. Both investigators contracted the disease, and the former had died.[12]

Reed decided to concentrate his research on Finlay's theory, after abandoning other widely-held current views on transmission. Bearing in mind the work of H R Carter (1852–1925) suggesting a five-day incubation period, and that a patient conveyed to a house to which a yellow fever 'victim' had been taken did

FIGURE 6.2 Walter Reed (1851–1902), leader of the successful American Yellow Fever Commission of 1900 (reproduced courtesy of The Wellcome Library, London).

FIGURE 6.3   James Carroll (1854–1907), a member of Reed's Commission (reproduced courtesy of The Wellcome Library, London).

not become infected until fifteen to twenty days after his arrival, and with Ross's recent work on the life-cycle of the malarial parasite in mind, Reed hypothesized that the infective organism developed *within* the mosquito. The first of Reed's team to receive the bite of an infected mosquito was Carroll, who was bitten on 27 August 1900 by a mosquito which had fed on a yellow fever 'victim' twelve days previously. Four days later Carroll was very ill, but he made a full recovery. A few days after this Lazear was bitten experimentally but did *not* develop yellow fever; however, on 13 September he was accidentally bitten (during a ward-round) by an *A. aegypti* mosquito; five days later he developed yellow fever and subsequently died. Thus, proof had emerged that the disease could be conveyed by an infected mosquito, although other means of transmission had not been ruled out.[13]

Reed next proceeded to construct a mosquito-proof hut in which seven beds were made up with sheets, etc., on which a yellow fever victim had lain.

JESSE W. LAZEAR, M.D.

FIGURE 6.4   Jesse Lazear (1866–1900), a member of Reed's Commission (reproduced courtesy of The Wellcome Library, London).

A doctor and six privates in the hospital corps duly occupied the beds and, after remaining in the hut for twenty days, none was clinically infected. In another hut, divided into two sections by a wire screen, fifteen infected mosquitoes were admitted into one part. Three non-immune American volunteers confined to the mosquito-free part did not develop yellow fever, but a volunteer in the mosquito compartment did develop the disease (from which he recovered) four days later. These experiments are reminiscent of the experiment carried out by Manson in the Roman Campagna (*see* Chapter 3).

The Commission thus concluded that *A. aegypti* is an intermediate host, and that the infection is contracted by non-immune individuals by the bite of a mosquito which has previously fed on infected blood. Furthermore, bites

FIGURE 6.5 Walter Myers (1872–1901), a Liverpool martyr of yellow fever research.

received more than twelve days after the mosquito has been infected neither cause the disease nor confer immunity. The disease is not transmitted by fomites; therefore disinfection is unnecessary. Finally, blood filtered through a bacterial filter remains infective.[14]

The Commission's findings, which became known in February 1901, were immediately applied to Havana, and were subsequently to have enormous impact on the construction of the Panama Canal.

Impact of Gorgas

W C Gorgas (1854–1920; Figure 6.6),[15] the son of a general in the US Army, was born in Alabama, graduated in New York in 1880, and immediately joined the Medical Department of the Army. In 1895, with the Spanish-American War almost over and the Spanish army back in Spain, he was sent to Havana as Chief

GENERAL W.C. GORGAS

FIGURE 6.6   W C Gorgas (1854–1920) (reproduced courtesy of The Wellcome Library, London).

Sanitary Officer. Believing at the time that yellow fever was a 'filth disease', he set out to improve the sanitary amenities of Havana; whereas the prevalence of malaria, typhoid and dysentery immediately fell, the introduction of non-immune immigrants from Spain in 1899 was associated with a rapid increase in mortality from yellow fever – so that by 1900 deaths from this disease were back to their pre-war level, when there were many non-immune soldiers around.

With the Reed Commission's findings publicized in early 1901 (see above), Gorgas realized that he must launch a specific campaign against *A. aegypti*, which breeds readily in domestic collections of water. By nursing all yellow fever patients in mosquito-proof rooms, and by making it a punishable offence to have mosquito larvae on a domestic premise, the mortality rate from yellow fever in Havana fell sharply.[16]

The Panama Canal

French work (headed by de Lesseps, who had previously constructed the Suez Canal) on construction of the Canal had started in 1881, but had been abandoned nine years later on account of enormous mortality from several disease entities, headed by yellow fever. In 1904 America launched another attempt at construction of the canal, and Gorgas was put in charge of sanitary measures; he introduced numerous draconian measures, as he had done at Havana. The disease rate (including that from yellow fever) thenceforth fell dramatically, and construction of the Canal was successfully completed in 1914.

Gorgas was later requested to investigate an outbreak of yellow fever in Senegal, but on his way there suffered a serious illness which was to prove fatal, and he died in London. Visited by King George V (1865–1936) in hospital, he was awarded the KCMG on his deathbed. His funeral service was held at St Paul's Cathedral, and a memorial service was subsequently held in Washington DC.[17]

YELLOW FEVER ELSEWHERE

Jungle yellow fever is transmitted by a different mosquito from that which spreads the *urban* variety; it has a reservoir in certain sub-human primates in Africa and South America. Therefore, a non-immune individual who has not received the vaccine is vulnerable.

In 1865 a minor outbreak occurred in Swansea, South Wales, which had been introduced by infected mosquitoes carried by and released from ships. As recently as 1930, a non-immune laboratory worker (who was investigating the disease in a sub-human primate) died of the disease at the Hospital for Tropical Diseases in London (see Figure 6.7).[18]

THE CAUSATIVE AGENT, AND SUCCESSFUL IMMUNIZATION

In 1886, Adolf Weil (1848–1916; see Chapter 18) had described what transpired to be the causative organism of leptospirosis, but which was considered for a time to be the cause of yellow fever. Various other bacteria had also from 1888 onwards been claimed to be the causative agent of yellow fever; in 1897 (the year prior to the beginning of the Reed Commission's work) an Italian bacteriologist suggested that *Bacillus icteroides* was responsible, and later, in 1919, a Japanese investigator, Hideyo Noguchi (1876–1928; see Chapter 18) incriminated a spirochaete – *Leptospira icteroides*. However, although this was considered for a time to be the causative agent in South America, this was discounted

FIGURE 6.7   The Hospital for Tropical Diseases in London as depicted in a picture postcard (*c*.1920).

in West Africa and, while researching the disease there, Noguchi unfortunately died of yellow fever. Another researcher to die of the disease in West Africa (at Yaba, Lagos, Nigeria, in September 1927) was Adrian Stokes (1887–1927) who, in experiments with sub-human primates, demonstrated that the causative agent was a filterable virus. The responsible virus was eventually identified by G W M Findlay (1893–1952) and J C Brown (1902–60).

In 1928, Edward Hindle (1886–1973) prepared the first attenuated vaccine against yellow fever. However, a satisfactory vaccine had to await the work of

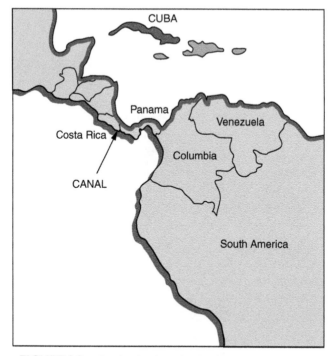

FIGURE 6.8   Map showing the major sites in the yellow fever saga.

Max Theiler (1899–1922) and his colleagues, in 1932 – one based on the 17D strain, a derivative of the Asibi strain, is used to this day.

Figure 6.8 shows the location of the major sites in the yellow fever saga.

NOTES

1   J K Fowler, W J Simpson, R Ross, W B Leishman (eds). *Yellow Fever Commission (West Africa): 2nd [British] Report* 1914: 141; C Lloyd, J L S Coulter. Yellow fever. In: *Medicine and the Navy 1200–1900*, Vol. 4. London, 1963: E & S Livingstone, pp. 183–94.

2   H Gardner (ed). *The New Oxford Book of English Verse 1250–1950*. Oxford, 1996: Oxford University Press, pp. 526–44.

3   J Beer. Coleridge, Samuel Taylor (1772–1834). In: H C G Matthew, B Harrison (eds), *Oxford Dictionary of National Biography*, Vol. 12. Oxford, 2004: Oxford University Press, pp. 572–90.

4   N G Goodman. *Benjamin Rush: physician and citizen 1746–1813*. Philadelphia, 1934: University of Pennsylvania Press, p. 421.

5   B Rush. *An Account of the Bilious remitting Yellow Fever, as it appeared in the City of Philadelphia in the year 1793*. Philadelphia, 1794: Thomas Dobson, p. 363; K J Bloom. *The Mississippi Valley's Great Yellow Fever Epidemic of 1878*. Baton Rouge, 1993: Louisiana State University Press, p. 290; L G Goodwin, C E G Smith. Yellow fever. In: F E G Cox (ed.), *The Wellcome Trust Illustrated history of Tropical Diseases*. London, 1996: The Wellcome Trust, pp. 142–7; C S Bryan, S W Moss, R J Kahn. Yellow fever in the Americas. *Infect Dis Clin North Am* 2004, 18: 275–92.

6   C Singer, E A Underwood. Yellow fever. In: *A Short History of Medicine*, 2nd edn. Oxford, 1962: Clarendon Press, pp. 466–81.

7   *Op cit.* See note 1 above (Lloyd, Coulter). See also G C Low. A retrospect of tropical medicine from 1894 to 1914. *Trans R Soc Trop Med Hyg* 1929, 23: 213–34.

8   H H Scott. Juan Carlos Finlay (1833–1915). In: *A History of Tropical Medicine*. London, 1939: Edward Arnold, pp. 1029–32; C E Finlay. *Carlos Finlay and Yellow Fever*. New York, 1940: Oxford University Press, p. 249.

9   *Op cit.* See note 6 above.

10 H H Scott. *A History of Tropical Medicine*, Vol. 2. London, 1939: Edward Arnold, pp. 1081–5; A E Traby. *Memoir of Walter Reed: the yellow fever episode*. London, 1943: Paul B Hoeber, Inc., p. 239; W B Bean. *Walter Reed: a biography*. Charlottesville, 1982: University Press of Virginia, p. 186. See also G C Cook. William Osler's fascination with diseases of warm climates. *J Med Biog* 1995, 3: 20–9. (Reed died suddenly of acute appendicitis in 1902.)

11 *Op cit.* See note 6 above; Sir William Osler was subsequently to give a tribute to Lazear during an address to the London School of Tropical Medicine. See also *op cit.* note 10 above (Cook) and (Scott), pp. 1056–7.

12 F Delaporte. The history of yellow fever: an essay on the birth of Tropical Medicine. London, 1991: MIT Press, p. 181.

13 *Op cit.* See note 6 above.

14 *Ibid.*

15 M D Gorgas, B J Hendrick. *William Crawford Gorgas: his life and work*. Philadelphia, 1924: Lea and Febiger, p. 359; F Martin. *Major General William Crawford Gorgas MC, USA*, 2nd edn. Chicago, 1929: Gorgas Memorial Institute, p. 74; J M Gibson. *Physician to the World: the life of General William C Gorgas*. London, 1989: University of Alabama Press, p. 315. See also *op cit.* See note 10 above (Scott), pp. 1037–55.

16 *Op cit.* See note 6 above.

17 *Ibid*; *op cit.* See note 10 above (Scott), pp. 1027–95.

18 C E G Smith, M E Gibson. Yellow fever in South Wales, 1865. *Med Hist* 1986, 30: 322–40; G C Cook. Fatal yellow fever contracted at the Hospital for Tropical Diseases, London, UK, in 1930. *Trans R Soc Trop Med Hyg* 1994, 88: 712–13. See also *op cit.* note 6 above.

7

James Cantlie (1851–1926): tropical surgeon, university administrator, and founder of the (Royal) Society of Tropical Medicine and Hygiene

The vast majority of pioneers of this discipline have been physicians, patholo-gists or microbiologists, or a combination of these. James Cantlie[1] FRCS (see Figure 7.1) is a notable exception.

Born in Banffshire, Scotland and educated at Botriphnie's School and Milne's Institution, Cantlie then entered Aberdeen University and in 1871 gradu-ated MA in natural science. After one year's training in medicine at Aberdeen, he transferred to the Charing Cross Hospital Medical School, London,[2] but returned to Aberdeen to graduate MB CM in 1873. Cantlie then returned to Charing Cross to take up a post as Instructor in Anatomy, and later Demonstrator in the same subject (1872–87); he was later House Physician, House Surgeon and Surgical Registrar. He became Assistant Surgeon in 1877 and Surgeon in 1886, but after two years he resigned from this position. Whilst at Charing Cross, Cantlie had taken a special interest in first aid and joined the newly-formed St John's Ambulance Association. Experience in the Charing Cross area also produced a longstanding interest in the urban poor. A lecture he gave in 1885 attributed the poor physique of Londoners to overcrowding and industrial pollution. This led to his portrayal by the media as a 'somewhat eccentric and even ridiculous figure'. In 1877 he was elected FRCS, and in 1882 he joined the London Scottish Volunteers as a surgeon; the following year he enrolled

FIGURE 7.1   Sir James Cantlie FRCS (reproduced courtesy of The Wellcome Library, London).

with the volunteer Medical Staff Corps, which had assumed much greater flexibility.

Involvement in tropical medicine

Late in 1877 Cantlie volunteered for service in Egypt, to offer assistance in a cholera epidemic which had been introduced by pilgrims from Mecca. This marked the beginning of his overseas work.

By now Cantlie's interest in medicine in the tropics was well established, and in 1887 he accepted an invitation from Patrick Manson to take over his practice in Hong Kong. Once there, he devoted himself to all aspects of medical work – including public health and smallpox vaccination. He also took a great interest in the distribution of leprosy in China and in other parts of the East Indies. His experience in an outbreak of bubonic plague in 1895 led to his appointment, when plague arrived in Bombay in 1896, as an adviser to the India Office.

However, Cantlie's greatest achievement in Hong Kong was a leading role in the establishment of a Medical College for Chinese students at the Alice Memorial Hospital, where he was, for seven years, the first Dean. One of the first graduates was Sun Yatsen, who in 1911 was to become the first President of the Chinese Republic.

London practice

In 1896, Cantlie returned to London to take the Chair of Applied Anatomy at Charing Cross Hospital. Shortly afterwards, Sun Yatsen was kidnapped and imprisoned in the Chinese legation in London; Cantlie was largely responsible for his safe release, and they thus remained good friends thereafter.

Cantlie developed a successful private practice in tropical medicine in London, and with Sir William Simpson (who had been a health officer in Calcutta and Editor of the *Indian Medical Gazette*) he launched the *Journal of Tropical Medicine and Hygiene* in 1898. He also assisted Manson (of whom he had been a staunch ally) in the establishment of the London School of Tropical Medicine (LSTM) at the Albert Dock Hospital (see Chapter 3), where he became the first Surgeon (he had a great interest in liver 'abscess') and a lecturer in tropical surgery. He was also responsible for a Tropical Section of the British Medical Association, which was inaugurated at the Edinburgh meeting in 1898; it was in fact at that meeting that Manson (President of the Section) read out the telegram from Ross announcing his elucidation of the avian malaria life-cycle. He was President of the Royal Society of Tropical Medicine (RSTMH), which he had founded jointly with Low (see below) in 1907, from 1921 until 1923.

Following his return to London, Cantlie's interest in the physical condition of the urban poor had been reawakened. Cantlie also maintained his interest in first aid and the organization of emergency medical aid; he became an adviser to the War Office on the establishment of the Territorial Force (later the Territorial Army) in 1907.

Later life

During the Great War (1914–18) Cantlie and his wife, who he had married in 1884, were greatly involved with the Red Cross and founded a College of Ambulance for those in distress due to disease or poverty.

Cantlie died at a nursing home in Dorset Square, London, on 28 May 1926, having been predeceased by his wife in 1921. He was buried at Cottered Cemetery, Buntingford, Hertfordshire.

One obituarist has written that he 'abounded in Scotch humour, and the faculty of imitation and the instincts of a born actor made him an admirable after-dinner speaker and singer'.[3] Another wrote that he 'combined Rabelaisian humour with the erudition of a Rabelair'. He also seems to have been an outstanding teacher.

President of the Royal Society of Tropical Medicine

Cantlie's Presidential Address to the RSTMH, given on 20 October 1921, was entitled 'Life Insurance in the Tropics'; his major theme was the extent to which tropical domicile affects the health of an individual, and whether it warrants an increased insurance premium. He certainly did not subscribe to the prevailing view that climate *per se* led to ill health; if malaria could be prevented (with mosquito nets and appropriate prophylaxis), he believed that the tropics were not in fact dangerous to the white man! He considered that the first three to five years (whilst acclimatization was taking place) were the most dangerous. Overall, he believed, rightly, that the dangers of life in the tropics had been grossly exaggerated. The incidence of amoebic dysentery, for example, had, since the introduction by Rogers (see Chapter 13) of emetine (a constituent of ipecacuanha), been greatly diminished. Cantlie also addressed 'the insurance of natives' (Hindoos, Malays, Chinamen and negros) for whom no statistics were in existence – a statement that was challenged in the subsequent discussion. He was of the opinion that only the 'anglicized' native, who assumed Western habits involving alcohol and diet, would require any insurance at all.

He knew of cases where Europeans had been recommended by their doctors to go to warm climates for health reasons, only to find that they had to pay increased insurance premiums. This was, he felt, a totally anomalous situation! Cantlie's overall message, then, was that insurance companies should be encouraged to reduce their premiums to those serving in the tropics.[4]

ROLE IN THE ORIGIN(S) AND EARLY DAYS OF THE (ROYAL) SOCIETY OF TROPICAL MEDICINE AND HYGIENE[5]

It is not widely known that Cantlie provided the initial stimulus for the foundation of this Society. He had in fact attempted to start a Society in 1899, but, owing to lack of support, this failed. In December 1906 he conveyed the idea to Low (see Chapter 8), who was immediately receptive. The two of them approached several other tropical physicians and surgeons, with variable results. Manson was 'at first doubtful about the idea, thinking that the time was not yet ripe;... he had heard that the RAMC were [contemplating] a similar organisation'. C W Daniels (see Chapter 19), medical tutor at the LSTM, was also 'somewhat guarded'. However, at a meeting held at the Colonial Office on 4 January, arranged by Cantlie and Low, it was 'unanimously decided to form a Society of Tropical Medicine and Hygiene'. The sub-committee formed at that meeting consisted of Manson, C N Melville, F M Sandwith, W Carnegie-Brown, 'a representative from Liverpool' and Cantlie, with Low as Secretary. This sub-committee subsequently met at Cantlie's house – 140 Harley Street – on 11, 16, 23 and 30 January 1907.[6]

A general meeting was then held at the Royal College of Physicians, in Pall Mall (see below), on 15 March, when the deliberations of the sub-committee were discussed; this meeting had been previously announced in the *Lancet* for 2 March:[7]

> The Society of Tropical Medicine and Hygiene was formed on Jan. 4th of this year at a meeting held at the Colonial Office for the purpose of the study and discussion of diseases met with in tropical countries. The second meeting of the society will be held on Friday, March 15th, at 5.30 p.m., at the Royal College of Physicians of London, to consider the rules drawn up by a sub-committee for the constitution of the society and the regulation of its work. So much has been done in recent years to advance the knowledge of the diseases of warm climates that there must be ample food for discussion and study in this branch of medicine and the meetings of the society should prove interesting not only to medical men with tropical experience but to every medical man engaged in the practice of medicine wherever his lot may be cast.

A further notice appeared in the *Lancet* for both 27 April and 4 May:[8]

> Society of Tropical Medicine and Hygiene. A meeting of the Society will be held at 5.30 p.m. on Friday, May 10th, 1907, at the rooms of the Royal Medico-Chirurgical Society (RMCS), 20, Hanover Square, London, W., to receive the report of the sub-committee appointed to draw up the rules and constitution, to elect members of Council, and to appoint the officers of the Society.

At a meeting on 10 May, Manson was elected first President, Ross (see Chapter 5) Vice-President, W Hartigan Treasurer, and F M Sandwith and W Carnegie-Brown joint Honorary Secretaries. The laws of the Society and the first Council were also agreed.[9]

The initial ordinary meeting of the Society took place at 20 Hanover Square (the headquarters of the Royal Medico-Chirurgical Society (RMCS), shortly to become the Royal Society of Medicine) in June 1907. It was on this occasion that Manson (see Chapter 3) delivered his inaugural Presidential Address. Many new Fellows were elected, including Professor Osler MD, FRS, 13 Norham Gardens, Oxford. There were few subsequent mentions of Osler's involvement with the newly formed Society, although he did open the discussion on a paper contributed by Ross on 15 February 1918 – less than two years before his death. Honorary Fellows elected at that inaugural meeting were:[10]

Professor Robert Koch
Professor Dr Paul Ehrlich
Professor A Laveran
Professor Raphael Blanchard
Lt Col W C Gorgas
Dr Theobald Smith
Professor Camillo Golgi
Professor Ettore Marchiafava
Professor Vasili Danilewski
Cathadratico Carlos Finlay
Jonathan Hutchinson
Professor Shibamiro Kitasato
Dr K Shiga.

FIGURE 7.2   The Royal College of Physicians, then situated at Pall Mall.

Early meetings of the Society were held at several venues: the Conference Hall of the Colonial Office; the RCP in Pall Mall (see Figure 7.2); the RMCS; 31 Cavendish Square; the Royal Devon and Exeter Hospital (Section of Tropical Diseases, BMA); 32 Harley Street; and 11 Chandos Street (Medical Society of London; see Figure 7.3). In subsequent years, the Society rented limited accommodation at the last of these venues.[11]

Plans for the fledgling Society to be subsumed by the Royal Society of Medicine (RSM)

Shortly after its foundation in 1907, attempts were made by the RSM (still situated at 20 Hanover Square, W1; see Figure 7.4) to incorporate the Society within its wide 'umbrella'. In June 1912 there were further attempts to amalgamate the Society with a tropical section of the RSM; the RSM's President from 1912 to 1914, Sir Francis Champneys Bt (1848–1920), clearly strongly favoured this strategy.[12]

The RSM had in fact come into being – its first Secretary was (Sir) John MacAlister (1856–1926)[13] – in 1907 at Hanover Square; it was not to move to 1 Wimpole Street until 1912. The RMCS merged with seventeen specialist societies (and it was hoped that this newly-formed Society would also join) in order to form the RSM.

FIGURE 7.3   The Medical Society of London, 11 Chandos Street, London W1.

FIGURE 7.4   20 Hanover Square: The Royal Medico-Chirurgical Society's headquarters, and the original home of the Royal Society of Medicine.

In November 1912, Sir William Leishman (see Chapter 12), at that time President of the Society, wrote to Ross indicating that he was glad that Ross had declined to accept the vice-presidency of the RSM's Tropical Section; he was clearly of the opinion that the RSM wanted to take over the Society's activities with a view to killing it! Furthermore, he felt the RSM had used the names of Manson and Cantlie in a strictly unauthorized way.[14] There was clearly, at that time, widespread interest in the possibility of the Society being absorbed by the RSM. In February 1913, the RSM laid down their 'rules' for amalgamation after consultation between Leishman and Champneys.[15] On 21 November 1913, it is clear that the Society again seriously discussed the possibility of amalgamation; however, a mere 60 Fellows were in favour, with 216 against! This decisive result was duly announced in both the *Lancet* and the *British Medical Journal*,[16] and the Society therefore maintained its independence![17]

Later controversial matters

In 1943, the Society was approached by the Colonial Office:[18]

A letter was read from Dr A G H Smart, Chief Medical Adviser to the Secretary of State for the Colonies, in which he asked 'whether it would be possible for the Colonial Office to join the [RSTMH] and receive some of the benefits of membership.

The Society discussed this suggestion and, perhaps not too surprisingly, rejected it.

Relationships with the Royal College of Physicians (RCP) seem to have been mildly strained in early days – for example, in 1918 the Council minutes refer to the revision of *Nomenclature of Diseases*, which apparently contained numerous errors in the entries on tropical disease(s). At the November meeting it was resolved to send a letter to the Registrar of the College:[19]

The Council of the Society of Tropical Medicine and Hygiene consider that it is regrettable that no authorities on Tropical Medicine were consulted on these, and other important points, before such a list was issued, and it trusts that when a further revision is contemplated this oversight will be remedied.

David Bruce
President
Society of Tropical Medicine and Hygiene

It seems, however, to have taken quite a while before this suggestion was followed up. It was not until 1944 that the Council Minutes record that:[20]

A letter was read from Lord Moran, President of the RCP, inviting the Council to appoint a delegate to serve on a Committee for the revision of 'The Nomenclature of Diseases'.

The matter of tropical consultants was also raised by the RCP in 1942:[21]

A letter was read from the [RCP] asking the Society to appoint an advisory committee of tropical consultants. After some discussion it was agreed that it was unnecessary to form a committee ... and that the [RCP] should be informed that the Council would at any time be willing to advise on any matter if desired ...

On 18 March 1921, more Honorary Fellows were nominated. However, Ross (see Chapter 5) took great exception to the inclusion of Professor Grassi's name; he was extremely critical of his scientific work and, in his usual paranoid way, considered that Grassi had attempted plagiarization of his major research findings. In support of his protest he submitted three letters, dated 1901, to Council, from Professor Robert Koch, A Laveran and Lord Lister, respectively. Professor Grassi was therefore not elected.[22]

The Society's Royal associations

In the early twentieth century tropical medicine had, for obvious reasons, become an important discipline in the life of the nation and also the Empire (see above). Therefore, the Society considered that it should seek to become a *Royal* Society. A letter addressed to the President, Sir William Simpson, in 1920, and preserved in the Society's archive, records:

HOME OFFICE
WHITEHALL, S.W.1.
9th June, 1920

Sir,

I am directed by the Secretary of State to inform you that he has laid before the King [George V] your application of the 13th April last for permission to use the prefix 'Royal' in the name of 'The Society of Tropical Medicine and Hygiene' and that His Majesty has been graciously pleased to command that the Society shall henceforth be known as 'The Royal Society of Tropical Medicine and Hygiene'.

I am, Sir
Your obedient Servant
A. J. Eagleston

This event, together with the early history of the Society, was recorded in *The Times* (17 June 1920):[23]

THE KING'S INTEREST

The King has been graciously pleased to command that the Society of Tropical Medicine and Hygiene shall henceforth be known as 'The *Royal* [my italics] Society of Tropical Medicine and Hygiene'.

  The formation of a society to promote the study of the diseases and hygiene of warm climates was considered by a number of interested persons in January, 1907, at a meeting at the Colonial Office. In due course the Society was constituted, with Sir Patrick Manson as President, and Major Ronald Ross as Vice-President. The first ordinary meeting was held in June, 1907, and since that date, except during and subsequent to the war, when the number of meetings was reduced, eight meetings have been held yearly and 13 volumes of 'transactions' have been published. The meetings are well attended, the number of Fellows present rarely falling below 50; on these occasions Fellows and their friends have the opportunity of exchanging views with workers from all parts of the world. At the end of the first year there were 186 Fellows and the number is now 660. In the last 12 months 175

Fellows have been elected. *From the first the society has been open to all persons interested in its objects, whether medical men or not, who are approved by the Council, and many veterinary practitioners are included among its Fellows* [my italics].

The Presidents in succession of Sir Patrick Manson have been Sir Ronald Ross, Sir William Leishman, Sir Havelock Charles, the late Dr F M Sandwith, and Sir David Bruce. The presidential chair is now occupied by Professor W J Simpson.

A Royal patron

In 1923, in response to an invitation from the Society, the then President (Surgeon Rear-Admiral Sir Percy W Bassett-Smith) received the following letter (which is also retained in the Society's archive):

> PRIVY PURSE OFFICE
> BUCKINGHAM PALACE, S.W.
> 11th December 1923
>
> Dear Sir,
>
> I have submitted your letter of the 8th inst., to the King, and, in reply, am commanded to inform you that His Majesty is graciously pleased to become Patron of the Royal Society of Tropical Medicine and Hygiene.
>
> Yours faithfully,
> F. H. Ponsonby [*sic*]
> Keeper of the Privy Purse

Contacts with the Patron seem to have remained close; in 1928, for example, the honorary secretaries of the Society received the following letter (also in the Society's archive):

> BUCKINGHAM PALACE
> 15th December, 1928
>
> Dear Sir,
>
> I am desired by the Queen [Mary] to thank the Council of the Royal Society of Tropical Medicine and Hygiene for their kind message of sympathy and good wishes for the King's speedy recovery, which you sent from their general meeting.
>
> Yours very truly,
> Clive Wigram

King George V, a heavy smoker, had become seriously ill with a chest infection and septicaemia on 21 November 1928. By 11 December he was barely conscious; however, on the following day his personal physician (later Lord Dawson of Penn) successfully aspirated a right-sided empyema. Following this he made a partial recovery which was punctuated by several relapses, the last of which (on 15 July 1929) took place a week after a thanksgiving service for his recovery at Westminster Abbey.[24]

CONCLUSION

James Cantlie had therefore contributed an enormous amount both to 'medicine in the tropics' (at Hong Kong) and also to the formal discipline by founding the (Royal) Society of Tropical Medicine and Hygiene, which celebrated its fiftieth anniversary in 1957, and its centenary in 2007.[25]

NOTES

1 Anonymous. Sir James Cantlie. *Lancet* 1926, i: 1121–2; Anonymous. Sir James Cantlie. *Br Med J* 1926, i: 1971–2; Anonymous. The late Sir James Cantlie. *Times, Lond* 1926, 31 May; Anonymous. Cantlie, Sir James (1861–1926). *Plarr's Lives: London: Royal College of Surgeons of England* 1930, 1: 192–3; N Cantlie, G Seaver. *Sir James Cantlie: a romance in medicine.* London, 1939: John Murray, p. 279; P Manson-Bahr. Sir James Cantlie, KBE, FRCS 1851–1926. In: *History of the School of Tropical Medicine in London (1899–1949).* London, 1956: H K Lewis, pp. 129–32; J C Stewart. *The Quality of Mercy: the lives of Sir James and Lady Cantlie.* London, 1983: George Allen and Unwin, p. 277; M Harrison. Cantlie, Sir James (1851–1926). In: H C G Matthew, B Harrison (eds), *Oxford Dictionary of National Biography*, Vol. 9. Oxford, 2004: Oxford University Press, pp. 962–4.

2 W Hunter. *Historical Account of Charing Cross Hospital and Medical School (University of London): original plan and statutes , rise and progress: founded in 1818.* London, 1914: John Murray, p. 309.

3 *Op cit.* See note 1 above (Cantlie, Seaver; Stewart).

4 J Cantlie. Life insurance in the tropics. *Trans R Soc Trop Med Hyg* 1921, 15: 109–16.

5 G C Low. The history of the foundation of the Society of Tropical Medicine and Hygiene. *Trans R Soc Trop Med* 1928, 22: 197–202.

6 *Ibid.*

7 *Ibid.*

8 *Ibid.*

9 Anonymous. *Lancet* 1907, i: 605.

10 Minutes of the Society of Tropical Medicine & Hygiene 1907, 26 June; Anonymous. *Lancet* 1907, i: 1583–4.

11 *Ibid*: 1907, 15 March; 1908, 18 December.

12 *Ibid*: 1912, 21 June, 23 September, 18 October.

13 Sir John MacAlister was a librarian, who subsequently became Secretary of the Royal Society of Medicine; this organization consisted of the amalgamation of seventeen separate societies. See also G C Cook. *John MacAlister's Other Vision: a history of the Fellowship of Postgraduate Medicine.* Oxford, 2005: Radcliffe Publishing, p. 178.

14 *Op cit.* See note 10 above, 1912, 21 June, 23 September, 18 October.

15 *Ibid*: 1913, 18 February; Anonymous. The Society of Tropical Medicine and Hygiene. *Lancet* 1913, ii: 158–9; J B Smith. The Royal Society of Medicine and the Society of Tropical Medicine and Hygiene. *Lancet* 1913, ii: 773–4; Anonymous. The Society of Tropical Medicine and Hygiene. *Br Med J* 1913, ii: 144–5.

16 *Ibid.* 1913, 21 November; Anonymous. The Society of Tropical Medicine and Hygiene. *Lancet* 1913, ii: 1639; Anonymous. The Society of Tropical Medicine and Hygiene. *Br Med J* 1913, ii: 1508.

17 G C Cook. Evolution: the art of survival. *Trans R Soc Trop Med Hyg* 1994, 88: 4–18. See also *op cit.* See note 5 above.

18 *Op cit.* See note 10 above: 1943, 21 October.

19 *Ibid.* 1918, 15 November.

20 *Ibid*. 1944, 16 March.

21 *Ibid*. 1942, 23 July.

22 *Ibid*. 1921, 18 March.

23 *Times, Lond* 1920, 17 June: 12.

24 H Nicolson. *King George the Fifth: his life and Reign*. London, 1952: Constable; F Watson. *Dawson of Penn: a biography*. London, 1950: Chatto and Windus; D Todd. *The Life and Times of George V*. London, 1973: Weidenfeld & Nicolson, p. 234. See also G C Cook. The practice of euthanasia at the highest level of society: the Lords Dawson (1864–1945) and Horder (1871–1955). *J Med Biog* 2006, 14: 90–92.

25 *Op cit*. See note 16 above; Anonymous. Tropical Jubilee. *Br Med J* 1957, 21 November; Anonymous. Prince Philip at Medical Dinner. *Daily Telegraph* 1957, 13 November.

8

George Carmichael Low (1872–1952): an underrated pioneer, and contributor to the (Royal) Society of Tropical Medicine and Hygiene

In the annals of the history of tropical medicine, the physician G C Low (see Figure 8.1)[1] does not immediately spring to the mind of most observers as being one of the major pioneers of the specialty. However, he made numerous contributions which had a highly significant impact on the development of the discipline.

Low was born on 14 October 1872 at Monifieth, Forfarshire; like so many of the other pioneers of the discipline of Tropical Medicine he had his origins in Scotland. He had distinguished under- and post-graduate careers. After Madras College he graduated in Arts at St Andrew's University, and then went to Edinburgh University where he gained an MB with first class honours in 1897. He later proceeded to the MD with Gold Medal in 1910, and the Gold Medal in Tropical Medicine in 1912. Low's junior appointments were at the Edinburgh Royal Infirmary.

In November 1899, Low joined Manson – then at the height of his fame – at the newly established London School of Tropical Medicine (LTSM).[2] As described earlier (see Chapter 3), this School had been founded by Manson, then Medical Adviser to the Colonial Office, with very strong support from one of the most powerful politicians of the day – Joseph Chamberlain MP, Secretary of State for the Colonies.[3]

Low later received various academic honours, including the Straits Settlements Gold Medal of the University of Edinburgh (1912) and the Mary Kingsley Medal

FIGURE 8.1   George Carmichael Low (1872–1952) an underrated pioneer (reproduced courtesy of The Wellcome Library, London).

of the Liverpool School of Tropical Medicine (1929). He was elected Membre d'Honneur of the Société Belge de Médecine Tropicale, and also made a corresponding member of the Société de Pathologie Exotique (Paris).[4]

Low died in London on 31 July 1952. In an anonymous obituary, *The Times* recorded that 'Low was one of the young "doctor naturalists" who were brought forward by Manson'. Sir Neil Hamilton Fairley (see Chapter 15) considered him to be

> An accurate observer and methodical to an unusual degree. …A sound physician and clinical teacher [who] possessed a profound knowledge of Tropical Medicine in its geographical, biological and pathological aspects. … A man of great intellectual honesty, said what he thought and adopted a justifiably critical and somewhat conservative attitude to things medical which were 'not proved'.

In another anonymous obituary, the *British Medical Journal* considered that, after his appointment as superintendent of the LSTM, Low 'found himself chained to the Metropolis, spending the rest of his career in the service of the School and as Physician to the Hospital for Tropical Diseases'. Those words were probably written by Sir Philip Manson-Bahr (see Chapter 18).[5]

Low therefore contributed enormously to the speciality of tropical medicine, which was emerging rapidly at the turn of the century. Some of his early research work was clearly of a very high standard. He was an outstanding teacher and administrator who was in many respects the linchpin of the clinical discipline in London. His greatest contribution was arguably, however, to the (Royal) Society of Tropical Medicine and Hygiene. In addition to his enormous service to the Society, he rescued it financially in its early days, and donated both the rostrum and the gavel (in use until the latter days of the twentieth century) in December and June 1931 respectively. Low did not receive a civil award, and there are many now – including some within the discipline of tropical medicine – who are not even familiar with his name, let alone his achievements. Why was Low underrated both then and now? He certainly served in the 'shadow' of Manson, and this may not have helped! Unlike most pioneers in this discipline he spent only a very short time (some two to three years) serving in a tropical environment; most of the major figures had devoted the bulk of their careers to practice and research in a 'warm climate'. However, these facts are inadequate to explain why so little is generally known about Low.

Low's first research project

In late 1899, Manson sent Low to Heidelberg and Vienna in order to learn a new technique for sectioning mosquitoes in celloidin (paraffin being an unsuitable medium), using a slide microtome. During Low's absence Manson received a batch of *Culex fatigans*, which had been infected with *Filaria nocturna* or *F. (Wuchereria) bancrofti* and preserved in alcohol, from Thomas Bancroft (son of Joseph) of Brisbane, Queensland. On returning to London, Low was given the task of sectioning these, and on 24 March 1900 he demonstrated filariae in the entire length of the proboscis sheath and emerging at the tip (see Figure 8.2).[6] It became clear, therefore, that man is almost certainly infected with larval filariae via the mosquito bite. Human infection had hitherto been considered to result from the ingestion of mosquito-contaminated water. (Although he had clearly demonstrated the man–mosquito portion of the life-cycle of *F. sanguinis hominis* (*F. bancrofti)* whilst working at Amoy, China, Manson had until then subscribed to the waterborne theory of human infection.) Low's work was published in the *British Medical Journal* for 16 June 1900. Shortly afterwards this discovery was confirmed by S P James,[7] working at Travancore, India.

At this early stage in his tropical medicine career, therefore, Low had delineated, for the first time, the completed life-cycle of a vector-borne helminthic (nematode) infection.

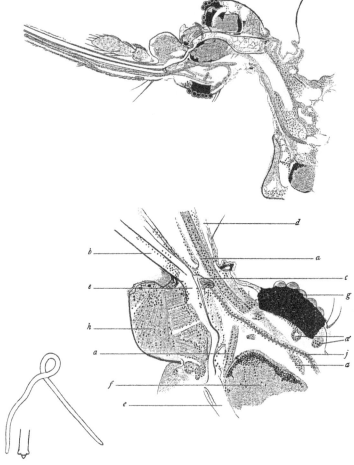

FIGURE 8.2 The proboscis sheath of *Culex fatigans* showing *Filaria nocturna*. G C Low *Br Med J* 1900, i: 1456–7.

The Roman Campagna expedition

In the autumn of 1900, Low, together with Dr L W Sambon, Signor Terzie (an artist) and a servant, travelled to a point at Ostia on the Roman Campagna about seven kilometres from Rome where *Plasmodium vivax* malaria was endemic.[8] During a three-month period the party confined themselves to a mosquito-proof hut (see Figure 8.3) between dusk and dawn, whilst living completely normal lives during the day; none of them experienced any of the clinical symptoms of malaria infection.[9]

In the second part of their investigation which was carried out with full support of the Italian malariologists *before* the priority controversy (see Chapter 5), an arrangement was made to send mosquitoes infected with *P. vivax* in a box

FIGURE 8.3 Mosquito-proof hut used in the Roman Campagna expedition 1900.

FIGURE 8.4 Mosquito box, used to convey malaria-infected mosquitoes from the Roman Campagna to London, preserved in the Library, London School of Hygiene & Tropical Medicine.

(transported in the diplomatic bag; see Figure 8.4) to London; here they were allowed to feed on Patrick Thurburn Manson, eldest son of Manson, and then a medical student at Guy's Hospital. He duly developed the symptoms of *P. vivax* malaria, which responded to quinine,[10] and there were two subsequent relapses the following year, both of which responded to chemotherapy.

Ross, working in India, had by this time produced a series of research papers indicating the mosquito transmission of avian malaria (see Chapter 5).[11] This work was communicated, on behalf of Ross, to the British Medical Association at the Edinburgh meeting in 1898 by Manson himself.[12] However, clear proof that *human* infection was a direct result of a mosquito bite was not at this time available, despite the fact that transmission by this route must, by then, have seemed virtually certain. The results of the Campagna expedition, although not accepted by the Italian malariologists (see Chapters 4 and 5), therefore constituted the first clear scientific evidence of this fact.

Later in 1900, evidence of yellow fever transmission to man via the mosquito bite (see Chapter 6) also became available.[13] Within this very short time, therefore, mosquito transmission of a viral, protozoan and helminthic infection had been scientifically documented; Low must surely have been at the forefront of medical/scientific investigators at that time.

The West Indies expedition

In 1901, Low, now the holder of a Cragg's Research Scholarship, set out to study helminthic disease in the Caribbean, and in particular the Windward Islands. With great industry and enormous energy, he travelled between and worked in St Lucia (Castries), Barbados, St Vincent, Trinidad, Grenada, St Kitts, British Guyana and Demerara. During this time he sent 32 letters to Manson (12 January 1901–2 April 1902), which have survived. They contain graphic accounts of the many and varied observations which he carried out during this period; many were epidemiological, but he also spent a great deal of time sectioning mosquitoes in an attempt to delineate the life-cycles of parasites. *Filaria demarquayi* and *F. ozzardi*[14] were subjects of particular interest. (Looked at in retrospect, it was, of course, unfortunate that he spent so much time working on two filarial species which are of very little clinical importance to *Homo sapiens*; however, he was able to accumulate a vast amount of scientific data.) He also confirmed the causative association between *F. bancroft* and elephantiasis, demonstrated by Manson at Amoy, China, some years previously (see Chapter 3). Another major project was to assist in the elimination of malaria from the Caribbean islands by draining swamps and areas containing stagnant water. It seems, however, that in this he was not taken seriously; this is made clear in a contemporary report in a local newspaper, the *Voice of St Lucia*:

> We hear that Dr Low the mosquito expert, has made a report on the malarial and mosquito-producing conditions of this island. The report will no doubt add to our stock of scientific knowledge. But we cannot help thinking that for all practical purposes the bliss of ignorance was in this matter decidedly preferable to the folly of wisdom.

Low sent this newspaper cutting to Manson, and was obviously considerably upset by the arrogance of the reporter; he again referred to this incident in his later correspondence. Low also mentioned in his letters to Manson that he had corresponded with Ross on the methods and practicability of eradicating malaria from various Caribbean islands.

The Royal Society's first sleeping sickness expedition to Uganda

Shortly after he had returned to England in 1902, Manson forwarded Low's name as leader of the first Royal Society expedition to attempt to find a solution to the problem of the 'negro lethargy' (sleeping sickness) which was sweeping the northern shores of Lake Victoria in 1901–2[15] (see Figure 8.5). Other members of the team were Aldo Castellani (see Chapter 14), an Italian graduate who had a background in bacteriology, and Cuthbert Christy (1863–1932), who had received training in epidemiology; he had been assistant to Haffkine in the Plague Laboratory at Bombay. The part to be undertaken by each member of the team was clearly demarcated prior to leaving England: Low was to investigate the association with *Filaria perstans*, which Manson had demonstrated in two cases of 'negro lethargy' in London, Castellani was to search for a bacterium, and Christy was to carry out epidemiological research into the epidemic. There were widespread fears that this disease (and it is perhaps no exaggeration to equate it in the minds of many with present-day AIDS) would spread to the 'Jewel in the Crown' – India.[16]

FIGURE 8.5   Map showing the extent of the outbreak of 'Negro lethargy' on the northern shore of Lake Victoria Nyanza (Royal Society, 1902).

Trypanosmiasis

Facts about the morphology of trypanosomes had of course been known long before the association with human disease was recognized. Gabriel Valentin (1810–83) had demonstrated a trypanosome in a salmon at Berne in 1841. Demonstration in other animals, including frogs, followed, and the generic name *trypanosome* was conferred on them by David Gruby (1810–98). The first demonstration in the blood of a mammal – the rat (*T. lewisi*) – had been by T R Lewis (1841–86) in 1878. Two years later, an army veterinary surgeon, Griffith Evans (1835–1935), described *T. evansi* in the blood of horses and camels suffering from *Surra*. The first authentic report of sleeping sickness is attributed to Atkins in 1735, and a clinical account was given by T M Winterbottom (1765–1859; see Chapter 1).

During the last decades of the nineteenth century, the disease had been well recognized by missionaries in various parts of Africa. In 1901 the 'negro lethargy' struck the northern shore of Lake Victoria Nyanza, and it was estimated that 20 000 individuals died or were beyond recovery. Low's small party (sponsored by the Royal Society) set out to investigate the cause. Manson was of the opinion that *Filaria (Mansonella) perstans* was the likely cause;[17] a further possibility seriously considered was that a streptococcus might be aetiologically involved. (Although a trypanosome had already been demonstrated by J E Dutton in 'trypanosoma fever' contracted in West Africa in 1901 (see Chapter 11), that disease and Ugandan sleeping sickness were not considered to be the same entity.[18])

The team sailed to Mombasa, Kenya, from whence they travelled by rail to the northern shores of Lake Victoria Nyanza. Here, Low and Castellani established a laboratory at Entebbe. According to one report, relationships between Low and Christy were already considerably strained by the time they reached Mombasa; it is likely that blows were exchanged at this time.[19] Relationships went from bad to worse, and Christy was to play only a minor role in the scientific aspects of the expedition (although, to be fair, he did subsequently publish a short account of the epidemiological aspects of the Ugandan epidemic).[20] Instead, he spent a great deal of his time trekking and big-game shooting. A report of the expedition's findings was subsequently published by the Royal Society.[21]

During this expedition Low wrote eight letters to Manson in London (24 June 1902–15 October 1902), which remain extant, describing local conditions and the progress of research. In one of them, dated 4 July 1902, Low expressed gratitude to Manson for the receipt of a Portuguese report outlining some newly acquired data on the possible aetiology of sleeping sickness; he mentioned that Castellani was especially interested in the information. (It seems certain that the organism under suspicion in this report was a streptococcus.) In Low's last letter to Manson written from Entebbe (15 October 1902), he conveyed the news that Castellani had detected a streptococcus-like organism in all cases of sleeping sickness which he had examined, making this the likely aetiological candidate. Shortly after this event (on 12 November 1902), Castellani saw and recorded a *Trypanosoma* sp. in cerebrospinal fluid from a case of sleeping sickness.[22] However, he discarded this observation, and persisted doggedly with the streptococcal theory (see also Chapter 14); it was not until David Bruce and D N Nabarro (see Chapter 9) arrived at Entebbe in 1903 that the aetiological significance of *Trypanosoma* sp. in this disease became clear.

In retrospect, Low's expedition seems to have been a failure. However, it is not so widely appreciated that it came extremely close to great distinction. Had the true significance of *Trypanosoma* sp. been appreciated by Castellani, and had he not been 'blinkered' by the streptococcal theory (probably influenced by the Portuguese report as well as his bacteriological training), the aetiology of this disease – considered by Manson the 'exemplar of a tropical disease' – would have been identified and Low and Castellani would have immediately joined those who had made major discoveries in tropical medicine.

Return to London

On return to London, Low was appointed superintendent of the London School of Tropical Medicine in late 1902, a post which he held from 1903 to 1905. He was to remain at the LSTM in various capacities until 1919 – as assistant physician (1912), physician (1918) and senior physician (1919). During this period he also became pathologist at the West London Hospital (1906–9), lecturer/consulting pathologist at the LSTM (1910–12) and assistant director of the Tropical Diseases Bureau (1912). Furthermore, he served as a lecturer at the Postgraduate College, King's College Hospital and the Westminster Hospital. During the Great War (1914–18) Low was appointed Major, Indian Medical Service; he served throughout at the ADH.

The Hospital and School moved to Endsleigh Gardens, London WC1, in 1920,[23] and Low was appointed senior physician to the Hospital for Tropical Diseases (HTD) – a post which he held until 1937. When the London School of Hygiene and Tropical Medicine (LSHTM) was opened in 1929, Low became director of the Division of Clinical Tropical Medicine, and remained in that capacity until 1937. Low's colleagues at this time included: Dr (later Sir) Philip Manson-Bahr, Dr (later Sir) Neil Hamilton Fairley and Sir Leonard Rogers. Throughout this period, Low made about 250 contributions to the literature on clinical aspects of tropical medicine and pathology – including filariasias (*F. bancrofti*, Loa loa and Onchocerca volvulus), schistosomiasis and other helminthic infections, malaria and blackwater fever, amoebiasis, liver abscess, yellow fever, leishmaniasis, trypanosomiasis, Malta fever, rat-bite fever, climatic buboes and ulcerating granuloma. He was responsible for recording one of the largest series of cases of tropical sprue (150), managed at the HTD.[24] He also wrote the section on tropical diseases in *Price's Textbook of Medicine* (with Hamilton Fairley), and that on amoebiasis (with Clifford Dobell) in *Byam and Archibald's Practice of Medicine in the Tropics*.

LOW AND THE ORIGIN(S) OF A PERMANENT 'HOME' FOR THE ROYAL SOCIETY OF TROPICAL MEDICINE AND HYGIENE – MANSON HOUSE

In its early days, the Society of Tropical Medicine (which had been founded in 1907; see Chapter 7) was a peripatetic organization. By 1920, the possibility of a permanent base was frequently under discussion. On 11 June 1920, Council seriously considered the advisability of adopting 23 Endsleigh Gardens (then owned by the Seamen's Hospital Society) as its Headquarters;[25] it would therefore be situated 'under the same roof as other institutes in London concerned with *Tropical Medicine*'. However, after a great deal of discussion it was resolved, on 28 June, to 'remain at Chandos Street till a suitable house can be found...'.[26]

By the 1920s the Society had largely recovered from the setbacks brought about by the Great War (1914–18), during which the total number of Fellows fell significantly. Despite earlier views (see Chapter 7), a closer relationship with the Royal Society of Medicine (RSM) was suggested by some – for example, on 18 June 1920:[27]

> Dr Harford [had] made a statement on his views with regard to the relation between the Royal Society of Tropical Medicine and Hygiene and the Royal Society of Medicine. He thought there should be co-ordination between the two Societies, having regard to the fact that tropical medicine was only a part of medicine in the tropics.

Accommodation at 11 Chandos Street was now proving inadequate, and it was decided that a suitable property, to be named Manson House, should be sought. In 1928 a subscription list was opened, and two years later a general appeal was launched. At a Council meeting on 11 December 1930, the following was minuted:[28]

> In connection with *Manson House* [my italics] the President [Dr G C Low] informed the Council that 26 Portland Place [London] was again in the market at a price of £22000. The house had been inspected by members of the House Sub-Committee and they unanimously recommended that if possible the property should be acquired for the Society even at the price asked ... . After a discussion it was decided that an effort should be made to secure the property for the Society ....

The appeal was by no means confined to London; according to the Royal Society of Tropical Medicine and Hygiene archive at the Wellcome Library, London, notices also appeared in the *Lourenço Marques Guardian*, *West Africa*, *The Pioneer of Allahabad*, *African World*, the *Tanganyika Opinion*, the *Tanganyika Standard*, the *Mombasa Times*, the *Tanganyika Times*, the *Hong Kong Telegraph*, the *Daily Chronicle of Delhi*, the *Star of Johannesburg* and the *Natal Mercury*, amongst others. At home, the *Yorkshire Herald*, *Country Life*, the *Aberdeen Press*, the *Wakefield Express* and *Farmer's Weekly* also ran appeals.

History of Manson House

The house had been designed by Robert Adam in 1778,[29] and in 1781 it was occupied by Lord Stormont. Shortly afterwards, it was the scene of a significant historical event: on 26 November, Lord Germain (1716–85), Commissioner of Trade and Plantations and Secretary of State for the Colonies (1775–82), accompanied by his Secretary, Lord Walsingham, arrived with 'stop press' news of General Cornwallis's surrender on 19 October at Yorktown, Virginia.[30] After collecting the Lord Chancellor, they called on the Prime Minister, Lord North, to break the news to him. Following this, they all proceeded to inform King George III of this memorable episode in the world's history.[31] Lord North and George III were subsequently blamed, as is well known, for the loss of the American colonies.

Following alterations to the property, including construction of an auditorium (see Figure 8.6) and the addition of furnishings, the total cost amounted to £29 822 – a sum not cleared until 18 May 1945. The first Council Meeting in Manson House took place on 16 July 1931, and the first Ordinary Meeting on 21 January of the following year. At a Council Meeting held in February 1932, 'It was announced that H.R.H. The Prince of Wales had very kindly consented to open Manson House on 17th March at 5.30. The President [G C Low] expressed his appreciation of the help rendered by Sir Austen Chamberlain and Sir Herbert Read [see Chapter 20] in approaching the Prince …'.[32]

FIGURE 8.6   (a)–(f) Views (both exterior and interior), of the newly-refurbished Manson House in 1931 (RSTMH archive).

FIGURE 8.6 (c)–(f)

Opening ceremony

On 17 March the opening ceremony duly took place (see Figure 8.7), with 250 Fellows seated in the auditorium.[33] The President delivered an address of welcome to His Royal Highness; following this, the Prince declared the house open.

FIGURE 8.7 Official opening ceremony at Manson House, 17 March 1932 (reproduced courtesy of the Royal Society of Tropical Medicine archive).

A vote of thanks was proposed by Sir Austen Chamberlain and seconded by W P MacArthur. Seated behind the rostrum were Sir Austen Chamberlain (son of Joseph), Sir Herbert Read (of the Colonial Office) and Lord Dawson of Penn.

In 1914, Dawson had been appointed Physician in Ordinary to King George V (Patron of the Society since 1911, who had granted the royal title in 1920). To many, though, he is perhaps best remembered for the bulletin concerning the King's illness which he composed and issued at 9.25 pm on 20 January 1936: 'The King's life is moving peacefully towards its close'. (In 1986, perusal of Lord Dawson's personal papers, which had recently been released, revealed that late on the evening of 20 January he injected morphine (grain 3/4) and cocaine (grain i) into the King's distended jugular vein.[34] The time of death was subsequently recorded as 11.35 pm, and the news was broadcast by the BBC at 12.10 am on 21 January. This manoeuvre was apparently carried out in order that the King's obituary notice would appear in the following morning's *Times*, rather than the evening papers – of which Dawson did not, it seems, entirely approve! This must be one of the most famous recorded cases of euthanasia, and in recounting these events *The Times* reminded its readership of a popular jingle of the time: 'Lord Dawson of Penn/Has killed many men/That is why we sing/"God Save The King."'[35])

Low served two terms as President in order to see the move to Manson House successfully completed. The only other President to have served for two terms was Sir Rickard Christophers (1873–1978), who continued in office for four years during the Second World War (1939–45) – during which period the Society suffered a further setback in numbers of Fellows, unpaid subscriptions, etc.

Manson House was damaged by enemy action during the war years:

[It] was first damaged (by blast from a bomb in Portland Place opposite the Langham Hotel) [and following this] it was agreed that urgent repairs to the roof and windows ... ought to be carried out at once ... [the property] was again damaged on April 16th 1941, this time by blast from bombs in Hallam Street behind the B.B.C. ...

In July 1942, a minute also records: 'the work required to make good all Air Raid Damage to Manson House ... amounts to £3,266.16.8 and includes a sum of £569.19.4 for temporary repairs which have now been completed...'.[36]

The later history of Manson House has also been recorded. In 1995 The Princess Royal officially opened the newly refurbished auditorium, renamed after G C Low.[37]

LOW THE ORNITHOLOGIST

Low was one of the most distinguished ornithologists of his day, and subsequently became Official Bird Watcher to Kensington Gardens. He was also often to be seen bird-watching at Staines, various London reservoirs, and on the Thames embankment. He frequented the Bird Room of the Natural History Museum, South Kensington, and became, successively, Secretary, Treasurer, and Editor of the British Ornithologists' Club. He was a Council Member of the Society for the Protection of Birds. As a Council Member of the Zoological Society of London, he was responsible for the centenary volume on *Aves* published in 1929. He also published *Literature of the Charadriiformes* (second edition) in 1932.

Whilst carrying out his duties at Kensington Gardens, Low must frequently have passed the imposing Speke Memorial, which commemorates the fact that the explorer had reached Victoria Nyanza in 1862 – 40 years before Low's unsuccessful sleeping sickness expedition to the same location. He must surely have felt a sense of nostalgia and disappointment that he and Castellani had so narrowly failed to solve the mystery of the 'negro lethargy' – an accomplishment which would without doubt have put him amongst the pioneers of tropical medicine as early as 1902.

LOW'S PRESIDENTIAL ADDRESS TO THE ROYAL SOCIETY OF TROPICAL MEDICINE AND HYGIENE

Low has left two immensely valuable historical accounts – his Presidential Address of 1929 (see Appendix 1), and also that of the previous year.[38] In the former he outlined the development of the discipline between 1894 and 1914, and in the latter he described the early years of the (Royal) Society of Tropical Medicine and Hygiene. He chose 1894 for the origin of the discipline, for it was in that year that Manson began developing his mosquito malaria hypothesis, and

Manson also met Ross for the first time on 9 April 1894. Low concentrated on these two decades for three reasons:

1. This was a 'period wherein all the isolated efforts [of the pioneers in the tropics] became welded together'
2. 'Some of the most remarkable discoveries [in the aetiology and treatment of disease] ever made in medicine [took place in these years]'
3. 'Tropical Medicine ... became [during these years] definitely defined as a special branch [of medicine]'.

Following a description of the establishment of the two Schools, at Liverpool and London (see also Chapter 20), he indicated that several others followed soon afterwards – in Paris, Hamburg, Belgium (1900) and, later, the USA. The history of researches as they stood in 1929 in filariasis (1899–1900), yellow fever (1900–1), trypanosomiasis (1901–3), ankylostomiasis (1901–2), Kala-azar (1903–4), undulant fever (1904), schistosomiasis (1904–7), plague (1907) and beri-beri (1907–9) followed.

In this far-reaching lecture, Low also alluded to the value of intravenous antimony (1906–1914) and emetine (1911–12) in various parasitoses, and outlined (as he had done in the past) the foundation of the (Royal) Society of Tropical Medicine and Hygiene in 1907. This society, he maintained, 'now [in 1929] holds a unique position as a centre or union for qualified workers in tropical medicine and allied sciences throughout the world'. He continued: 'I am certain that in the future the Society will stand as the head of Tropical Medicine for Britain, and one might say for the whole world'. The appeal for funds for the purchase of Manson House was, he indicated, progressing well, because it was a 'necessity that the Society should have its own home' – a fact which has sadly, in recent years, been lost!

NOTES

1 Anonymous. Obituary: George Carmichael Low. *Br Med J* 1952, ii: 341–2; Anonymous. Low, George Carmichael. *Munk's Roll*, Vol. 4. London: Royal College of Physicians, pp. 594–5; N H Fairley. Obituary: George Carmichael Low. *Trans R Soc Trop Med Hyg* 1952, 46: 571–3; Anonymous. Obituary: George Carmichael Low. *Lancet* 1952, ii: 296–7; Anonymous. G Carmichael Low. *Times, Lond* 1952, 1 August: 8; G C Cook. George Carmichael Low FRCP: twelfth President of the Society and underrated pioneer of tropical medicine. *Trans R Soc Trop Med Hyg* 1993, 87: 355–60; G C Cook. George Carmichael Low FRCP: an underrated figure in British tropical medicine. *J R Coll Phys Lond* 1993, 27: 81–2; M Worboys. Low, George Carmichael (1872–1952). In: H C G Matthew, B Harrison (eds), *Oxford Dictionary of National Biography*, Vol. 34. Oxford, 2004: Oxford University Press, pp. 550–51.

2 G C Low. Presidential Address: a retrospect of tropical medicine from 1894 to 1914. *Trans R Soc Trop Med Hyg* 1929, 23: 213–34; G C Cook. *From the Greenwich Hulks to Old St Pancras: a history of tropical disease in London*. London, 1992: Athlone Press, p. 332.

3 *Op cit*. See note 2 above (Low).

4 *Op cit*. See note 2 above (Cook).

5 *Op cit*. See note 1 above.

6 G C Low. A recent observation on filaria nocturna in Culex: probably mode of infection of man. *Br Med J* 1900, i: 1456–7; Anonymous. Lymphatic filariasis: tropical medicine's origin will not go away. *Lancet* 1987, i: 1409–10; G C Cook. Discovery and clinical importance of the filariasis. *Infect Dis Clin North Am* 2004, 128: 29–30. See also *Op cit*, note 2 above (Low).

7 S P James. On the metamorphosis of the filarial sanguinis hominis in mosquitoes. *Br Med J* 1900, ii: 533–7.

8 *Op cit*. See note 2 above (Cook).

9 P Manson. Experimental proof of the mosquito-malaria theory. *Br Med J* 1900, ii: 149–51.

10 *Ibid*.

11 R Ross. Report on the cultivation of proteosoma, Labbé in grey mosquitoes. *Indian Med Gaz* 1898, 33: 401–8, 448–52.

12 Anonymous. The role of the mosquito in the evolution of the malarial parasite: the recent researches of Surgeon-Major Ronald Ross, IMS. *Lancet* 1898, ii: 488–9.

13 F Delaporte. *The History of Yellow Fever: An Essay on the Birth of Tropical Medicine*. London, 1991: MIT Press, p. 181; *Op cit*. See note 2 above (Low).

14 J J C Buckley. On the development, in *Culicoides furens* Poey, of *Filaria* (=*Mansonella*) *ozzardi* Manson 1897. *Journal of Helminthology* 1934, 12: 99–118; D I Grove. *A History of Human Helminthology*. Wallingford, 1990: CAB International.

15 G C Cook. Correspondence from Dr George Carmichael Low to Dr Patrick Manson during the first Ugandan sleeping sickness expedition. *J Med Biog* 1993, 1: 215–29.

16 M Lyons. *The Colonial Disease: A Social History of Sleeping Sickness in Northern Zaire, 1900–1940*. Cambridge, 1992: Cambridge University Press.

17 *Op cit*. See note 15 above.

18 J E Dutton. Preliminary note upon a trypanosome occurring in the blood of man. *Thompson Yates Laboratory Reports* 1902, 4: 455–67.

19 *Op cit*. See note 15 above. See also F L Salanha (ed.), *The Haffkine Institute 1899–1974*. Bombay, 1974: Government Central Press, p. 148.

20 C Christy. The distribution of sleeping sickness, filarial perstans, &c, in East Equatorial Africa. *Reports of the Sleeping Sickness Commission*, II (No. 3). London, 1903: Harrison and Sons, pp. 3–8.

21 *Ibid*. See also G C Low. A retrospect of tropical medicine from 1894 to 1914. *Trans R Soc Trop Med Hyg* 1929, 23: 213–32.

22 A Castellani. On the discovery of a species of trypanosome in the cerebro-spinal fluid of cases of sleeping sickness. *Proceedings of the Royal Society of London* 1903, 71: 501–8. See also: B I Williams. African trypanosomiasis. In: F E G Cox (ed.), *The Wellcome Trust Illustrated History of Tropical Diseases*. London, 1996: Wellcome Trust, pp. 178–91; F E G Cox. History of sleeping sickness (African trypanosomiasis). *Infect Dis Clin North Am* 2004, 18: 231–45.

23 *Op cit*. See note 2 above (Cook); Anonymous. Diseases of the tropics: hospital need of £35000: Duke of York's appeal. *Times, Lond* 1930, 9 April; Anonymous. Seamen's Hospital: the Duke of York's appeal for aid. *Lloyds List* 1930, 10 April.

24 G C Low. Sprue: an analytical study of 150 cases. *Q J Med* 1928, 21: 523–34.

25 RSTMH Council minutes, 1920, 11 June.

26 *Ibid*. 1920, 28 June.

27 *Ibid*. 1920, 18 June.

28 *Ibid*. 1930, 11 December. See also: Anonymous. In memory of Manson. *Times, Lond* 1929, 18 December; G C Low, R Ross, A Balfour, J W W Stephens, A G Bagshawe, C W Wenyon. Father of tropical medicine: memorial to Sir Patrick Manson. *Times, Lond* 1929, 18 December; Anonymous. Father of Tropical Medicine: some memories of Sir P Manson. *Observer* 1929, 22 December; Anonymous. Sir Patrick Manson. *Med Press Circ* 1929, 25 December; Anonymous. Father of Tropical Medicine: Sir A Chamberlain and a memorial. *Times, Lond* 1929, 28 December; E Brumpt. Father of Tropical Medicine. *Times, Lond* 1929, 31 December; F Fülleborn. Father of Tropical Medicine. *Times, Lond* 1930, 11 January; Anonymous. In memory of Manson. *Lancet* 1930, 18 January; Anonymous. In memory of Manson. *Br Med J* 1930, 18 January; Anonymous.

Memorial to the Father of Tropical Medicine. *Med Press Circ* 1930, 31 January; Anonymous. Patrick Manson. *Med J Aust* 1930, 1 March; Anonymous. No 26, Portland Place: purchase by Royal Society of Tropical Medicine. *Times, Lond* 1931, 28 March; Anonymous. Royal Society of Tropical Medicine and Hygiene: memorial to Sir Patrick Manson. *Med Press* 1931, 1 July.

29 'Royal Society of Tropical Medicine and Hygiene 1907–1957', RSTMH archive.

30 After France and Spain, with their powerful fleets, entered the War of American Independence (1775–83) against Britain, the Royal Navy briefly lost control of the seas. Cornwallis had retreated with his army to Yorktown, where he was secure provided he could receive supplies by sea. On 17 October 1781, a British rescue fleet set out from New York – but too late. Cornwallis was already outnumbered, outgunned, and rapidly running out of food. When the Royal Navy failed to come to his aid; he was forced to surrender his large army to the colonialists – a defeat which led to the British Government's abandonment of the struggle. Independence of the colonies was recognized at the 'Peace of Paris' in 1783, and they developed their own form of government, eventually uniting under the presidency of George Washington in 1787.

31 'Royal Society of Tropical Medicine and Hygiene 1907–1957', RSTMH archive.

32 *Op cit.* See note 25 above: 1932, 18 February.

33 Anonymous. Opening of Manson House: headquarters of the Royal Society of Tropical Medicine and Hygiene by His Royal Highness the Prince of Wales, KG, on Thursday March 17th 1932, at 5.30 pm. *Trans R Soc Trop Med Hyg* 1932, 25: 493–501. See also: Anonymous. Easy to treat: message to the King and Queen. *Times, Lond* 1930, 3 March; Anonymous. The Prince ill: slight attack of malaria: sudden return from hunting: hopes of recovery in few days. *Daily Mail* 1930, 3 March; *Op cit.* See note 1 above (Cook 1993).

34 F Watson. The death of George V. *History Today* 1986, 36: 21–30; G C Cook. The practice of euthanasia at the highest level of society: the Lords Dawson (1864–1945) and Horder (1871–1955). *J Med Biog* 2006, 14: 90–92.

35 Anonymous. The King's Peace? *Times, Lond* 1986, 28 November: 17.

36 RSTMH Executive Committee minutes 1941, 17 July; 1942, 23 July.

37 In 2003, Manson House, the greatest and by far the most important possession of the RSTMH, which had formed the focal point of the Society for more than 70 years, was hastily sold for £3.2 million. This deal was arranged by Council, with the President, A M Tomkins presiding, without adequate reference to Fellows, numerous throughout the world, whose predecessors had subscribed to its purchase as a memorial to the 'Father of Tropical Medicine'. The true market value of this building (arguably the most perfect Adam building in Portland Place) was, as that time, probably in the region of £10 million.

38 *Op cit.* See note 2 above (Low); see also Low, G C (1928). The history of the foundation of the Society of Tropical Medicine and Hygiene. *Trans R Soc Trop Med Hyg* 1928, 22: 197–202; P Manson-Bahr. *History of the School of Tropical Medicine in London (1899–1949)*. London, 1956: H K Lewis, pp. 158–62.

9

David Bruce (1855–1931): Malta fever, nagana, and East African trypanosomiasis

David Bruce (1855–1931; Figure 9.1) made significant contributions regarding brucellosis (in Malta), nagana (in Zululand), and east African trypanosomiasis (in Uganda). In all of these ventures he was accompanied by a fellow-worker – his wife, formerly Mary Elizabeth Steele (1849–1931).

Bruce was born in Melbourne, Australia, of Scottish parents. At six years of age he travelled to Scotland, and was educated at Stirling High School. At seventeen he joined the business world, but shortly afterwards entered Edinburgh University and graduated MB CM in 1881. As well as being a keen and accomplished sportsman (he excelled at boxing), he had an enthusiastic interest in natural history. Following a brief period in general practice he entered the Army Medical School at Netley, where his major interests were in pathology and research in that discipline.[1]

Malta fever (brucellosis)

Bruce's first assignment in the RAMC was at Valletta, Malta, in 1884. He had married in 1883, and his wife, a trained microbiologist, was also a most accomplished artist. Her accurate sketches and paintings were to enhance Bruce's outstanding research activities throughout his life (see Figure 9.2).

FIGURE 9.1 David Bruce (1855–1931) (reproduced courtesy of The Wellcome Library, London).

Malta was, of course, a British possession with strategically important naval and military bases. British soldiers based in Valletta, however, had a marked morbidity from Malta (Mediterranean) fever, which caused about 120000 missed days of work each year; the average time period off-duty was 90 days. There was also significant mortality. At the time, poor hygiene and contaminated water supplies were (as was the case with most other diseases) considered causative. Bruce demonstrated a Gram-negative micrococcus (*Micrococcus melitensis*), 3 μm in diameter, in specimens of spleen, kidney and liver obtained at post-mortem. Monkey inoculation of this micrococcus cultured on peptone broth produced a death sixteen days later in one of three experiments, following a febrile illness. Post-mortem there was congestion of liver and spleen, but no involvement of Peyer's patches – making typhoid fever most unlikely. Following this, seven other monkeys were inoculated; all developed a febrile illness and four died. A report incriminating this organism (subsequently named *Brucella melitensis*) was transmitted to the Pasteur Institute, Paris.

In 1904, a team consisting of Major Horricks, Staff Surgeon E A Shaw, Dr T Zammit, Captain I Crawford Kennedy RAMC and several others set out to ascertain the mode of transmission of this causative organism. Zammit discovered

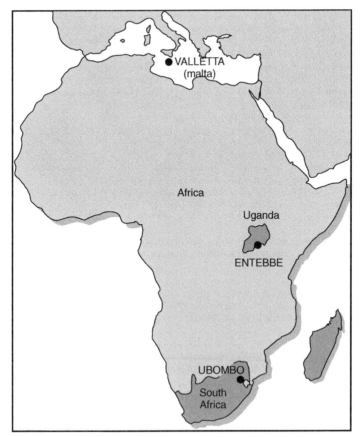

FIGURE 9.2 Map showing sites of Bruce's major researches.

the organism in the blood of a goat (see Figure 9.3). It was subsequently shown that more than 50 per cent of the island's goats were infected, most subclinically, and 11 per cent excreted the organism in their urine. Thus, the source of infection was identified. This fact was confirmed when Maltese goats were introduced into Rhodesia, when an epidemic of undulant fever resulted, and on the *SS Joshua Nicholson*, which was shipping goats to the USA, where everyone – including the captain – who drank goat's milk, suffered from the disease.

The part played by the Maltese bacteriologist Scicluna in unravelling this problem is difficult to determine.

Guiseppe Caruana Scicluna (1853–1921)

Scicluna had studied at the Pasteur Institute, Paris, and worked as a sanitary inspector/analytical chemist in the Police Department in Valletta. Later, he

FIGURE 9.3 Photograph, taken in Malta, depicting milking of a goat, the milk of which contained the causative organism of brucellosis (reproduced courtesy of The Wellcome Library, London).

was Superintendent of Public Health and Chief Government Medical Officer. Appreciation of his work later recorded:[2]

> His collaboration with Bruce in the isolation and cultivation of the specific bacterial cause of Undulant Fever – the micrococcus melitensis – from the spleen of patients, has never been widely enough known to connect his name with such an important discovery and to elicit recognition. The first spleen smears were made on agar plates and tubes prepared at the Public Health Department and were incubated in an improvised stove made out of a biscuit tin!

In 1887, he had apparently made several attempts to inoculate blood samples from the fingers of ten patients suffering from the disease into agar tubes. From one, colonies appeared in all the culture tubes incubated at 37°C for 68 hours. Following this success, Bruce isolated the micrococcus from biopsy tissues. John Eyre wrote that 'Bruce carried out his cultivation and inoculation experiments in the laboratory of the Public Analyst, Scicluna'.[3] However, Scicluna prepared the agar plates and succeeded in culturing the bacteria from the spleens of four British soldiers dying of the disease. Madkour, in a book published in 1959, commented that Bruce never acknowledged Scicluna's contributions in this research.[4] It also seems likely that Scicluna suggested to Zammit that goat's milk was the source of human infection.[5]

Scicluna thus probably played a crucial role in the solution of the aetiology of Malta fever; Figure 9.4 shows a plaque on the exterior wall of the laboratory in Valletta in which Bruce and Scicluna both worked.[6]

By eliminating goat's milk from the diet, the military authorities eliminated Malta fever from British troops.

This discovery led to Bruce being elected FRS, and he was also appointed Assistant Professor in Pathology at the RAM School at Netley; here he worked with Almroth Wright, who at that time occupied the Chair of Pathology.[7]

FIGURE 9.4   Plaque situated on the exterior wall of the laboratory in Valletta, Malta, where both Bruce and Scicluna worked.

Nagana

In 1894, Bruce was posted to Pietermaritzburg (Natal, South Africa) for military field service. At the time, cattle and also horses were dying in hundreds from a disease the origin(s) of which was obscure; the Zulus gave it the name *nagana*.[8] As a result, the country was faced with ruin, and grave economic stress affected the indigenous stock-raisers and white settlers.[9] On the initiative of the Governor of Natal and High Commissioner of Zululand, Sir Walter Hely-Hutchinson, who had recently served as Lieutenant-Governor of Malta whilst Bruce was working on the Malta (undulant) fever problem (see above), Bruce was ordered to proceed on secondment from Natal to Zululand in an attempt to discern the aetiology of this mysterious disease. Nagana and (tsetse) fly disease of travellers and hunters were, at that time, considered separate entities.[10]

Africans in certain parts of the continent were already aware that nagana resulted from a bite of the tsetse fly. The inhabitants of this region of Africa (and also West Africa, where it was called *surra*) and India had been seriously bothered by the disease for at least a century,[11] and perhaps since 1742.[12] David Livingstone (1813–73) was in fact familiar with it on the banks of the Zambesi in 1847;[13] he described the disease in 1857[14] and produced an accurate drawing of *Glossina morsitans*. Many local residents, some of European origin, were of the opinion that it was caused by the bite of the tsetse fly (which elaborated a poison within itself),[15] leading to the death of the animal some ten days after this event. An alternative theory, apparently held by most Zulus, was that game animals harboured the disease, and that it was transmitted to cattle in food and water contaminated by them – perhaps faecally.[16] We now know that both theories possessed an element of truth.[17] Livingstone was fully convinced that the disease did not affect man; he and his party lived in a heavily infected area for months without ill effect. This suggests that human strains of *Trypanosoma brucii* were not at that time present in this part of Africa.

Bruce and his wife[18] thus set off on a long trek to Ubombo Hill, some 650 metres (2000 feet) above sea level in northern Zululand, by mule and ox wagon, to take up their challenge on 27 October 1894.[19] However, from notes written at Pietermaritzburg before they set off it is clear that Bruce was already intellectually involved in the problem, for he made an attempt to infect animals by injecting into them watery and alcoholic tsetse-fly extracts.

Following his work which had unravelled the cause of Malta (undulant) fever, Bruce felt that a bacterium was the most likely candidate. Furthermore, he had spent the year 1888 at Robert Koch's institution in Berlin – 'a Mecca for budding bacteriologists'.[20] At Ubombo Hill, where they arrived on 24 November 1894, the Bruces took over a wattle-and-daub hut that had belonged to a 'squatter' married to a Zulu woman. Figure 9.5 shows Bruce at Ubombo while he was researching nagana.[21] His work began by undertaking daily blood examinations (including cell counts) on a 'brown cow'. Each examination revealed micrococci and slender, poorly-staining bacilli; however, on the sixth day he recorded in his notes 'also Haematozoa.'[22]

FIGURE 9.5 Bruce in Zululand while researching the cause of nagana (reproduced courtesy of The Wellcome Library, London).

Bruce had very little knowledge of trypanosomes, and 'at first thought that the wriggling object might possibly be a small filaria'.[23] The observation was repeated on a 'black and white cow'. Healthy calves were shown to develop nagana when taken down to the low country;[24] their blood also contained 'haematozoa' – in one case numbering 10000 per mm³. Whilst in the low country, two of the Bruces' dogs – pointers – were bitten by tsetse flies; shortly after returning to the hill, both developed acute, fatal nagana with 'haematozoa' in their blood. One of them, 'John Keats' apparently had four organisms ('wriggling about like little snakes') per erythrocyte. When the splenic pulp and blood of this dog were cultured, they were shown to be bacteriologically sterile. Bruce also examined large numbers of healthy cattle, and noted that 'haematozoa' were invariably absent. Within two months, and probably a mere five to six weeks[25] of his arrival in Zululand, Bruce seems to have clearly established a positive correlation between nagana and blood 'haematozoa', and furthermore that the 'organism' was responsible for both nagana and tsetse fly disease.[26] He recorded these observations in his *Preliminary Report on the Tsetse Fly Disease or Nagana, in Zululand*;[27] this was later expanded in his *Further Report on the Tsetse Fly Disease or Nagana, in Zululand*.[28] Later, in 1915, he described this work in considerable detail in four Croonian lectures delivered to the Royal College of Physicians.[29]

At this exciting point in the saga, Bruce was recalled, on 26 January 1895, to Natal.[30] However, on reaching Pietermaritzburg he discovered that there had been no compelling reason for this order and, following communications between the Governor and the War Office (who were seemingly pretty uncooperative), the Bruces were able to return to Ubombo on 8 September 1895. However, by the time they arrived, seven months had been wasted. Bruce was by this time fully convinced that the 'haematozoa' were causatively related to nagana, but this had to be confirmed scientifically. Therefore, he infected healthy animals by inoculating the blood of diseased ones. A minor setback came when a dog fed on coagulated, infected blood developed nagana – a surprise observation which seemed to support the 'Zulu theory' (see above); in retrospect, the only reasonable explanation must be that trypanosomes entered via an oral abrasion(s).[31]

When healthy cattle were taken to the low country, muzzled, and fed on fodder brought down from the hill, they still contracted nagana; those kept on the hill and fed herbage brought from the low country remained fit and well. In an epidemiological study carried out over several miles of country, he found that in some kraals the Zulus no longer kept cattle because they knew they would die of nagana, whereas in others they flourished; the former were, he noted, in bush (scrub), with game animals nearby, and the latter in open country. The role of game animals in relation to nagana remained unclear until Bruce succeeded in 'infecting' a dog by inoculating the blood of an antelope; he repeated this experiment many times, but results were inconsistent. He later concluded that about one-quarter of the local herbivorous game animals (tolerant to infection) harboured 'haematozoa' in their blood, and therefore served as reservoirs of infection.[32] He then proceeded to demonstrate, by inoculation experiments, that game animals can be infected, but remain asymptomatic.[33]

At an entomological level, Bruce also carried out tsetse-fly breeding experiments[34] and noted that larvae were retained in the abdomen of the parent fly. He showed that *G. palpalis* was responsible. He proceeded to demonstrate living 'haematozoa' in the proboscis of tsetse flies fed on infected animals.[35] One hour after feeding, they were present in the stomach in clumps of a dozen or so, and remained very active; after four hours, activity was undiminished. Bruce's notes recorded that the longest survival period of the organism(s) in the tsetse-fly intestine was 118 hours. From these experiments Bruce concluded that the fly can transmit the organisms for up to 48 hours after feeding on an infected animal, and that after 72 hours infection must be unlikely. In order to establish whether or not a later developmental cycle existed within the fly, he carefully dissected the gut of infected specimens; although he recorded 'disc-like bodies, spirilla, and so forth',[36] he decided that these were independent of the 'haematozoa'.

At this point, the South African war intervened and he was unable to keep the flies for a full three weeks in order to establish the developmental cycle within the insects' salivary glands. This was ultimately unravelled by F K Kleine (1861–1950) in 1908, more than a decade later, who demonstrated that the responsible trypanosome develops in the tsetse fly in a similar way to that of the malarial parasite and filarial sp. in the mosquito, and that infection is transferred in saliva when the fly bites. At the siege of Ladysmith (which lasted from October 1899 until March 1900), in a battle with the Boers, Bruce was compelled to work as a surgeon (his prime objective was to investigate an outbreak of enteric fever amongst British troops) while his wife was a sister with the Red Cross. Of 563 deaths during the siege, 393 resulted from typhoid. For their work there, Bruce both received a medal and was promoted in rank, while his wife was awarded the Royal Red Cross Medal.[37]

Bruce thus established beyond doubt, by a series of elegant experiments, that nagana is caused by a 'haematozoa', later named *Trypanosoma brucii (brucei)* by H G Plummer (1850–1918) and J R Bradford (1863–1935);[38] this is conveyed by an infected tsetse fly, which is itself infected by feeding on game animals, which form the major reservoir of infection. Bruce was therefore the first investigator to demonstrate transmission of a protozoan parasite by an insect bite;[39] he was also the first to demonstrate the developmental cycle within the tsetse fly. He satisfied himself, furthermore, that dosing with arsenic – first established to be efficacious by David Livingstone in a mare afflicted by nagana[40] – had an inhibitory effect on trypanosomes in the blood of animals, but did not prevent a subsequent infection.[41]

Bruce's researches that led to the solution of the *nagana* problem are regarded by most historians of medicine/science as his greatest contribution.[42] This was scientifically more challenging than the brucellosis, and east African trypanosomiasis research that was to follow.

East African trypanosomiasis

The first Royal Society sleeping sickness expedition (consisting of Low, Castellani and Christy; see Chapter 8), which set out to unravel the aetiology of the 'negro lethargy' which in 1901 was ravaging the local population on the northern shores of Lake Victoria Nyanza, failed to identify the cause of the disease. Bruce, with his research ability now firmly established (he had already solved the causes of brucellosis and nagana), was at this point sent by the Royal Society to determine the origin of yet another disease which was causing a great deal of local morbidity and mortality. He was accompanied in this venture by David Nabarro (1874–1958), described as a 'quiet Portuguese' pathologist, and, of course, his wife.

On 12 March 1903, the Bruces joined Castellani and Christy at Entebbe, Uganda (Low had already left for England). Castellani had by then demonstrated 'haematozoa' in the cerebrospinal fluid (CSF) of five cases (and in the peripheral blood of one), although, being a bacteriologist and highly impressed (and influenced) by a Portuguese report, implicated a diplo-coccus.

Bruce, largely as a result of his nagana work, felt it pertinent to research the possible aetiological role of the trypanosome. Using centrifuged specimens of CSF, he detected the protozoan in 70 per cent of 34 cases of 'negro lethargy', but in none of 12 controls. Subsequent work revealed trypanosomes in 100 per cent of 40 cases examined. Extrapolating, Bruce concluded that these cases and 'trypanosome fever' (already investigated in West Africa; see Chapter 11) represented different stages of a single disease.

Injection of 'infected' CSF into a monkey produced a sleeping sickness-like illness, but since the animal was found at post-mortem to have co-existent tuberculosis, this result was discounted. Subsequent inoculation of blood from a sleeping-sickness sufferer subcutaneously was followed by a similar illness, and at post-mortem trypanosomes could be demonstrated in the central nervous system. An epidemiological study throughout the area showed a close correlation between the proportion of *Glossina palpalis* infected and the prevalence of sleeping sickness. Extensive studies on animals revealed that dogs and rats are partially susceptible to sleeping sickness, but that guinea-pigs, donkeys, goats and sheep are refractory. A preliminary report was published in the *British Medical Journal*.[43]

On 25 May, Bruce and Nabarro were joined by E D W Greig (1874–1950) and they wrote a 'Further Report' which cleared up many outstanding problems. In 1909, Kleine (see above), an associate of Koch and a member of the German Sleeping Sickness Commission, showed that flies bred in the laboratory only become infective after about twenty days after their infecting feed; this was subsequently confirmed by Bruce and his co-workers.

On 28 August 1903, the Bruces left Africa with the intention of continuing their work on brucellosis in Malta. In 1908 Bruce was, however, appointed Director of the third Royal Society Commission on Sleeping Sickness. This was basically an epidemiological investigation, the results of which were reported in his Croonian Lectures for 1915.[44]

Bruce concluded that there were three distinct patterns of trypanosomal disease:

1. Nagana, which affected a vast region from Sudan to Zululand
2. *Trypanosoma rhodesiense*, responsible for sleeping sickness in Nyasaland (now Malawi) and Rhodesia (now Zambia and Zimbabwe) and transmitted by *Glossina morsitans*
3. *T gambiense*, responsible for 'trypanosoma fever' in the Congo and Uganda and transmitted by *Glossina palpalis*.

In 1910, J W W Stephens (1865–1946) and H B Fantham (1875–1937) in fact discovered *Trypanosoma rhodesiense* in Nyasaland (now Malawi) and northern Rhodesia (now Zambia); furthermore, Allan Kinghorn (1880–1955) and Warrington Yorke (1883–1943) confirmed that the transmitting fly was *G morsitans*.

BRUCE'S LATER LIFE

In 1914, the year of the outbreak of the Great War, Bruce was appointed Commandant of the Royal Army Medical College. Here, he reviewed the efficacy of typhoid and tetanus inoculation.[45] He continued in this capacity until his retirement in May 1919, when he was appointed a representative of the Royal Society on the governing body of the Lister Institute.

In his latter years, Bruce received numerous honours. He was a staunch advocate of prevention, as this quotation from his Presidential Address to the British Association for the Advancement of Science shows:[46]

> It must be no longer said that the man was so sick that he had to send for the doctor. The medical practitioner of the future must examine the man while he is apparently well, to detect any incipient departure from normal and to teach and urge modes of living comfortable to the laws of personal health, and the public health authorities that man's environment is in accordance with scientific teaching.

BRUCE'S PRESIDENTIAL ADDRESS TO THE SOCIETY OF TROPICAL MEDICINE

Bruce did not speak, surprisingly, about any of his major discoveries – the causes of brucellosis, nagana or east African trypanosomiasis – in this address, but instead he analysed 1000 cases of tetanus; he was still (in 1917) Chairman of the War Office Committee for the Study of Tetanus. At the beginning of his lecture, he apologized to his audience for the fact that he was not addressing a strictly *tropical* disease, but one which 'in a time of war [the Great War was still in progress] has an important place among war diseases'.

Sir David Bruce died on 27 November 1931 while the funeral service for his wife was in progress at Christ Church, Westminster.[47]

NOTES

1 Anonymous. Sir David Bruce: discoveries in tropical disease. *Times, Lond* 1931, 28 November; Anonymous. Sir David Bruce, KCB, DSc, LLD, MB, FRCP, FRS, Major-General AMS (retd). *Br Med J* 1931, ii: 1067–9; Anonymous. Sir David Bruce, KCB, MD Edin, FRS, Major-General, AMS (retd). *Lancet* 1931, ii: 1270–71; Anonymous. Major-General Sir David Bruce, KCB, LL D, DSc, FRCP, FRS. *J RAMC* 1932, 58: 1–4; A E Hamerton. Major-General Sir David Bruce, KCB, DSc, LL D, FRCP, FRS, late AMS. *Trans R Soc Trop Med Hyg* 1932, 25: 305–12; J R B (probably John Rose Bradford). Sir David Bruce – 1855–1931. *Obituary Notices of Fellows of the Royal Society 1932–1935*, Vol. 1. London, 1932: Royal Society, pp. 79–85; M Robertson. Sir David Bruce: an appreciation of the man and his work. *J RAMC* 1955, 101, 91–9; W J Tulloch. Sir David Bruce: an appreciation. *J RAMC* 1955, 101: 81–90; E E Vella. Major-General Sir David Bruce, KCB. *J RAMC* 1973, 119: 131–44; Anonymous. *Munk's Roll*, Vol. 4. London: Royal College of Physicians, pp. 515–17; H H Scott. *A History of Tropical Medicine*. London, 1939: Arnold, pp. 1018–21; S R Christophers, H J Power. Bruce, Sir David (1855–1931). In: H C G Matthew, B Harrison (eds), *Oxford Dictionary of National Biography*, Vol. 8. Oxford, 2004: Oxford University Press, pp. 287–9. See also D Doyle. Eponymous doctors associated with Edinburgh, Part 2 – David Bruce, John Cheyne, William Stokes, Alexander Munro *Secundus*, Joseph Gamgee. *J R Coll Phys Edinb* 2006, 36: 374–81.

2 H V Wyatt. Dr G Caruana Scicluna, the first Maltese microbiologist. *J Med Biog* 2000, 8: 191–3.

3 J H Eyre. The Milroy Lectures on Melitensis Septicaemia (Malta or Mediterranean Fever). Lecture III. *Lancet* 1908, i: 1826–32.

4 M M Madkour. Historical aspects of brucellosis. In: M M Madkour (ed.), *Brucellosis*. London, 1989: Butterworth, pp. 1–10.

5 *Op cit.* See note 1 above.

6 *Ibid.*

7 G C Low. A retrospect of Tropical Medicine from 1894 to 1914. *Trans R Soc Trop Med Hyg* 1929, 23: 213–32; H H Scott. Undulant fever. In: *A History of Tropical Medicine*. London, 1929: Arnold, pp. 768–80; B J S Grogono. Sir David and Lady Bruce. A superb combination in the elucidation and prevention of devastating diseases. Part II: Further adventures and triumphs. *J Med Biog* 1995, 3: 79–83, 125–32. See also F E G Cox. Brucellosis. In: F E G Cox (ed.), *The Wellcome Trust Illustrated History of Tropical Diseases*. London, 1996: Wellcome Trust, pp. 50–59.

8 *Op cit.* See note 2 above; W MacArthur. An account of some of Sir David Bruce's researches, based on his own manuscript notes. *Trans R Soc Trop Med Hyg* 1955, 49: 404–12.

9 *Op cit.* See note 1 above (Tulloch).

10 *Op cit.* See note 1 above (Anonymous 1932, JRB); G C Low. A retrospect of Tropical Medicine from 1894 to 1914. *Trans R Soc Trop Med Hyg* 1929, 23: 213–32.

11 C Singer, E A Underwood. *A Short History of Medicine*, 2nd edition. Oxford, 1962: Clarendon Press, pp. 481–7. See also *Op cit.* See note 2 above.

12 *Op cit.* See note 1 above (Anonymous, *Times, Lond* 1931).

13 D Livingstone. Arsenic as a remedy for the tsetse bite. *Lancet* 1858, i: 360–61.

14 D Livingstone. *A Popular Account of Missionary Travels and Researches in South Africa*. London, 1857: John Murray, pp. 55–8.

15 *Op cit.* See notes 3, and 1 (Anonymous, *Times, Lond* 1931) above. See also D Bruce *Preliminary Report on the Tsetse Fly Disease or Nagana in Zululand*. Durban, 1895: Bennet and Davis, p. 28.

16 *Op cit.* See note 3 above.

17 *Op cit.* See note 1 above (Robertson).

18 *Op cit.* See note 1 above (Tulloch).

19 *Op cit.* See note 15 above (Bruce).

20 *Op cit.* See note 1 above (JRB, Vella).

21 J J Joubert, C H J Schutte, D J Irons, P J Fripp. Ubombo and the site of David Bruce's discovery of *Trypanosoma brucei*. *Trans R Soc Trop Med Hyg* 1993, 87: 494–5.

22 *Op cit.* See note 3 above.

23 *Op cit.* See note 15 above (Bruce).

24 *Op cit.* See note 1 above (Robertson).

25 *Op cit.* See note 1 above (Tulloch).

26 *Op cit.* See note 1 above (Anonymous 1932, Hamerton, Robertson, Tulloch, Vella).

27 *Op cit.* See note 15 above (Bruce).

28 D Bruce. *Further Report on the Tsetse Fly Disease or Nagana in Zululand.* London, 1896: Harrison, p. 69.

29 *Op cit.* See note 15 above (Bruce).

30 *Op cit.* See notes 3 and 15 above (Bruce).

31 *Op cit.* See note 3 above.

32 *Ibid.*

33 *Op cit.* See note 1 above (Anonymous, *Times, Lond* 1931).

34 *Ibid.* (Robertson).

35 *Op cit.* See note 3 above.

36 *Ibid.*

37 *Op cit.* See note 7 above (Grogono).

38 H G Plimmer, J R Bradford. A preliminary note on the morphology and distribution of the organism found in the tsetse fly disease. *Proceedings of the Royal Society of London* 1899, 65: 274–81.

39 G C Cook. Some early British contributions to tropical medicine. *J Infect* 1993, 27: 325–33. See also *op cit.* note 1 above (Anonymous 1932, Vella).

40 *Op cit.* See note 8 above.

41 *Op cit.* See note 3 above.

42 G C Cook. Sir David Bruce's elucidation of the aetiology of *nagana* – exactly one hundred years ago. *Trans R Soc Trop Med Hyg* 1994, 88: 257–8. See also *op cit.* See note 2 above.

43 D Bruce, D Nabarro, E D W Greig. Further report on sleeping sickness in Uganda. *Royal Society: Reports of The Sleeping Sickness Commission*, Vol. 4. London, 1903: Harrison & Sons, pp. 1–87; D Bruce, A Nabarro, E D W Greig. Etiology of sleeping sickness. *Br Med J* 1903, ii: 1343–50; Anonymous. Sleeping sickness. *Br Med J* 1903, ii: 1351–2; *op cit.* See note 5 above (Low); G C Cook. Sir David Bruce's research on trypanosomes. *J Med Biog* 1996, 4: 61; B I Williams. African trypanosomiasis. In: F E G Cox (ed.), *The Wellcome Trust Illustrated History of Tropical Diseases.* London, 1996: Wellcome Trust, pp. 178–91.

44 D Bruce. The Croonian Lectures on Trypanosomes causing Disease in Man and Domestic Animals in Central Africa. *Lancet* 1915, i: 1073–8; ii: 1–10, 48–53, 91–7. See also F E G Cox. History of sleeping sickness (African trypanosomiasis). *Infect Dis Clin North Am* 2004, 18: 231–45.

45 *Op cit.* See note 7 above (Grogono).

46 D Bruce. The prevention of disease. *Science* 1924, 60: 109–24.

47 *Op cit.* See note 1 above.

10

The schistosomiasis saga: Theodor Bilharz (1825–62), Robert Leiper (1881–1969), and the Japanese investigators

In his classic book entitled *A History of Tropical Medicine* (1939), Scott gave only scant attention to schistosomiasis; Singer and Underwood (1962), although dealing with the history of the major tropical infections, entirely omitted reference to this group of diseases (see Chapter 15), which are estimated to affect 200 million individuals globally, and with an increasing prevalence.[1]

URINARY SCHISTOSOMIASIS

Historically, this infection in Africa probably originated when hunter-gatherers settled into and established agricultural communities – beginning around the great lakes and spreading along the trade routes into the Nile valley. The ancient Egyptians certainly suffered from *Schistosoma haematobium* infection. European armies, including that of Napoleon at Aboukir Bay, suffered from haematuria, as did British troops in the Boer War (1890–1902) and Australian solders at Tel el Kebir in Egypt in the Great War (1914–18).[2]

Causative agent(s)

The helminth responsible for this infection was discovered by a young German doctor, Theodor Bilharz (1825–62; Figure 10.1),[3] while performing post-mortems

FIGURE 10.1   Theodor Bilharz (1825–62), a German investigator, who discovered the causative organism of *Schistosoma haematobrium* at Cairo in 1851 (reproduced courtesy of The Wellcome Library, London).

at Kesr el Aini hospital, Cairo, in 1851. He reported his finding(s) to Professor Carl von Siebold (1804–85; Figure 10.2) in Breslau.[4]

Transmission

Arthur Looss (1861–1923; Figure 10.3), P Sousino (1835–1901) and T S Cobbold (1828–86) all attempted, without success, to infect local snails with miracidia hatched from the eggs of infected individuals. All were familiar with

Dᵣ von SIEBOLD

*Director des Grosherz Hess Medicinal Collegiums
and 1ᵉ Physicats Arzt zu Darmstadt*

FIGURE 10.2   Carl von Siebold (1804–85), who confirmed Bilharz's discovery of the cause of urinary schistosomiasis in 1851 (reproduced courtesy of The Wellcome Library, London).

the life-cycles of other trematode worms that develop in snails. The first of these investigators held that the miracidia infected man directly.

In 1915, the Royal Army Medical Corps, at Lord Kitchener's direction, sent Robert Leiper (1881–1969; Figure 10.4)[5] and a team to Egypt. They managed to infect a rat and a mouse with cercariae. Leiper also demonstrated the life-cycle – miracidia, sporocyst, daughter sporocyst, cercariae – within the snail. These results were subsequently confirmed by Manson-Bahr and Fairley in Imperial troops of the Egyptian expeditionary forces in 1916–18.[6]

INTESTINAL SCHISTOSOMIASIS

J Harley, in South Africa in 1864, became convinced that some patients with endemic haematuria were also infected with an organism, the eggs of which had a lateral rather than a terminal spine. It soon became clear that they were associated

Arthur Looss.

FIGURE 10.3   Arthur Looss (1861–1923), who also later made major contributions to ankylostomiasis research (reproduced courtesy of The Wellcome Library, London).

with intestinal rather than urinary disease. In 1903, Manson himself described a patient from the West Indies with lateral-spined ova in his stool, and also suggested two separate species, one producing bladder and the other rectal disease. Louis Sambon (1866–1931), who was at that time working with Manson at the London School of Tropical Medicine, named the new species *S. mansoni* in 1907. However, whether they were in fact two separate species was a highly contentious matter until Leiper and his team demonstrated in 1915 (in Cairo) that these distinct miracidia were in fact attracted by two separate species of snail – those with terminal spines to *Bulinus*, and those with lateral spines to *Biomphalaria*. Leiper concluded that the two kinds of egg were 'characteristic products of two distinct species … and are spread by different intermediary hosts'.[7]

RESEARCHES ON SOUTH-EAST ASIAN SCHISTOSOMIASIS

In 1847, D Fujii visited the Kawanami district of Japan. Individuals exposed to water in the rice-fields developed pruritic papules on their legs, and cattle and

FIGURE 10.4   Robert Thomson Leiper, FRS (1881–1969), who was responsible for demonstrating the complete life-cycles of *S. haematobium* and *S. mansoni* (reproduced courtesy of The Wellcome Library, London).

horses were also affected. Fujii wrote a report (which did not come to light in the Western world until 1909) in which he described a syndrome consisting of 'fever', pallor, bloody/mucoid diarrhoea, muscle-wasting and abdominal swelling. Although this is probably the first report of a *S. japonicum* infection, it represents a disease of great antiquity – recently, 2000-year-old corpses have been shown to contain ova of *S. japonicum*.[8]

The eggs of *S. japonicum* had been described in 1888, but the source was not located. In 1904, Katsurada discovered the eggs (devoid of spines) in the Yamanashi Prefecture. In the same year, A Fujinami (1870–1934) found both eggs and adult worms in a patient with the Katayama syndrome (a form of *acute* schistosomiasis in the non-immune individual, consisting of fever, an urticarial rash, and hepato-splenomegaly). A claim that John Catto had first described the female trematode containing *S. japonicum* eggs in 1905 was negated because they were identical to Fujinami's findings, of 1904.[9]

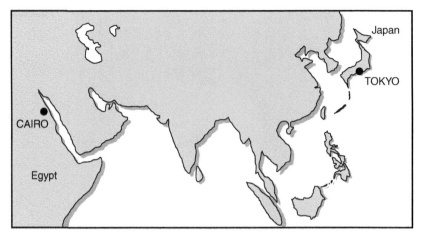

FIGURE 10.5 Map showing the sites at which major discoveries in the elucidation of *Schistosoma* spp. infection were made.

It rapidly transpired that this is a zoonotic species and therefore there are several animal 'models'. Thus, transmission was rapidly worked out. Fujinami and Nakamura demonstrated in 1909 that infection occurred through intact skin, and by 1914 the entire life-cycle had been unravelled.[10]

It should be noted that most of this work had been published in Japanese, and therefore little was known about *Schistosoma japonicum* in the West. Leiper and E Atkinson travelled to Japan to carry out work on this helminthiasis, only to discover that most of the work they intended carrying out had already been accomplished by Japanese physicians and parasitologists.[11]

Figure 10.5 shows the sites of the major discoveries in the schistosomiasis saga.

MANAGEMENT

The traditional anthelminthics – potassium iodide, male fern extract (*Dryopteris felix mas*), santonin and quassia – have little beneficial action against schistosomes. It was John Christopherson who pioneered intravenous tartar emetic in Sudan in 1918, although priority was later challenged unsuccessfully by J E R McDonagh (1881–1965). Christopherson was actually treating a patient suffering from *kala-azar*, who had a co-existent schistosomal infection. For many years intravenous trivalent antimony (potassium or sodium antimonyl tartrate) remained the standard treatment, despite significant toxic effects, and the occasional fatality. Less toxic derivatives (stibophen, lithium, antimony thiomalate, sodium antimonyl dimer-capro succinate and sodium antimonyl gluconate) were introduced but were less potent, especially against *S. mansoni*. Emetine was effective, but was also toxic.

Really effective and relatively safe anti-schistosome compounds were not introduced until the Second World War (1939–45) in Germany; these, derived from thioxanthene, had the advantage of being administered orally. It was not until 1972 that an oral compound effective against all three major species of schistosomes – praziquantal – was introduced.[12]

Prevention

Attempts to abolish the intermediate host – the various species of fresh-water snail – have invariably failed. Copper sulphate, sodium pentachloraphenate and niclosamide have all been tried as molluscicides. Hygiene programmes, aimed at the abolition of urination and defecation in infected rivers and lakes, have also failed in most locations, although relative success has been reported from China.

Relative immunity occurs in *Homo sapiens*, and although work on a potential vaccine is in progress, this, like a malaria vaccine, seems at the present far over the horizon!

ANCYLOSTOMIASIS

Arthur Looss (Figure 10.3) was also an important pioneer in elucidating the life-cycle of the hookworm (both *Ancylostoma duodenale* and *Necator americanus*). The disease(s) associated with this helminth were probably recognized in Egyptian times, and the adult worm was visualized by Angelo Dubini (1813–1902) in 1838; it was soon shown that the presence of adults in the upper small-intestine could be confirmed by the presence of ova in the faeces.

In 1880, hypochromic anaemia, caused by this helminth, was common in miners constructing the St Gothard tunnel. In 1898, Looss showed that infection occurred not by the oral route, but via intact skin. The complete life-cycle was documented in a monograph (1905–11). His interest began when he accidentally inoculated himself with a drop of culture-fluid, which produced local irritation ('ground itch').[13]

This infection, by impairment of working ability, was subsequently the focus of a campaign by the Rockefeller Foundation. The initial object was to eliminate the infection from the southern states of the USA (see also Chapter 17).

NOTES

1 H H Scott. *A History of Tropical Medicine*. London, 1939: Arnold. C Singer, E A Underwood. *A Short History of Medicine*, 2nd edn. Oxford, 1962: Clarendon Press, p. 854. See also A A F Mahmoud. Schistosomiasis (bilharziasis): from antiquity to the present. *Infect Dis Clin North Am* 2004, 18: 207–18.

2 T Bilharz, C T von Siebold. Ein Bietrag zur Helminthographia humana, aus brieflichen nittheilungen del Dr Bilharz in Cairo, nebst bermerkungen von Prof C The von Siebold in Breslau. *Zeitschift fur Wissenchatilich Zoologie* 1852, 4: 53–75; L G Goodwin. In: F E G Cox (ed.), *The Wellcome Trust Illustrated History of Tropical Diseases*. London, 1996: Wellcome Trust, pp. 264–73.

3 Theodor Maximilian Bilharz (1825–62) was a German Professor of Zoology. Working in Cairo, he discovered the adult helminth *Schistosoma haematobium* in portal systemic blood in 1851. It was named after him by T S Cobbold (1828–86).

4 F G Cawston. Some notes on the differentiation of closely-allied xchistosomes. *Parasitology* 1922, 12: 245–7; *Ibid.* Distribution of snails serving as intermediate hosts of flukes in South Africa, *Parasitology* 1924, 16: 67–8.

5 Robert Thomson Leiper FRCP, FRS (1881–1969) was born at Kilmarnock, Ayrshire, and educated at Warwick School, Mason College Birmingham, and Glasgow University, graduating MB, ChB in 1904. He joined Manson as a helminthologist at the London School of Tropical Medicine in 1905, and remained there or at the London School of Hygiene and Tropical Medicine until 1946. He subsequently worked in Scotland, south-east Asia, east and west Africa, Egypt, the Caribbean and South America. His major interests were in the parasites of man and domestic animals. During the Second World War (1939–45) he served as a Lieutenant Colonel in the RAMC, and was posted to Cairo to investigate schistosomiasis in British troops; he had previously worked there with Looss in 1906–7. He established the complete lifecycles of *Schistosoma haematobium, S. mansoni*, dracontiasis, and *Loa loa*. Leiper was also involved in many helminthological journals. Later in life he received numerous honours. See also: P C C Garnham. Leiper, Robert Thomson. *Munk's Roll*, Vol. 6. London, Royal College of Physicians, pp. 279–81; P Manson-Bahr. In: *History of the School of Tropical Medicine in London (1899–1949)*. London, 1956: H K Lewis, pp. 227–9; Anonymous. *Times, Lond* 1969, 23 May; Anonymous. *Lancet* 1969, i: 1103–4; Anonymous. *Br Med J* 1969, ii: 579–80; P C C Garnham. Robert Thomson Leiper 1881–1969. In: *Biographical Memoirs of Fellows of the Royal Society*. London, 1970: The Royal Society, pp. 385–404; J Farley. Leiper, Robert Thomson (1881–1969). In: H C G Matthew, B Harrison (eds), *Oxford Dictionary of National Biography*, Vol. 33. Oxford, 2004: Oxford University Press, pp. 280–81.

6 A Looss. Wüurmer und die ihnen hervorgerufenen Erkrankungen. In: Mense (ed.), *Handbuch der Tropenkrankheiten*, Vol. 2, 2nd edn. Leipzig, 1914: J A Barth, pp. 331–74; R T Leiper. Report on the results of the *Bilharzia* mission in Egypt. *J RAMC* 1915, 25: 1–55, 147–92, 253–67; P Manson-Bahr, N H Fairley. Observations on bilharziasis amongst the Egyptian Expeditionary Force. *Parasitology* 1920, 12: 33–71.

7 J Harley. On the endemic haematuria of the Cape of Good Hope. *Med-Chir Trans* 1864, 47: 55–72 (see also *Lancet* 1864, i: 156–7); P Sonsino. Discovery of the life history of *Bilharzia haematobia* (Cobbold). *Lancet* 1893, ii: 621–2; P Manson. *Tropical Diseases. A manual of the diseases of warm climates*, 3rd edn. London, 1903: Cassell; L W Sambon: What is '*Schistosoma mansoni*'? London, 1907: Sambon. *J Trop Med Hyg* 1909, 12: 1–11; A Looss. What is *Schistosoma mansoni* Sambon 1907? *Ann Trop Med Parasitol* 1908, 2: 153–91; R T Leiper. On the relation between the terminal-spined and lateral-spined eggs of *Bilharzia*. *Br Med J* 1916, i: 411; R T Leiper. Report on the results of the *Bilharzia* mission in Egypt 1915. *J RAMC* 1918, 30: 253–60; G C Low. A retrospect of Tropical Medicine from 1894 to 1914. *Trans R Soc Trop Med Hyg* 1929, 23: 213–32.

8 D Fujii. An account of a journey to Karayoma. *Chugan Iji Shinpo* 1909, 691: 55–6; Mao Shou-Pai, Shao Bao-Ruo. Schistosomiasis control in the People's Republic of China. *Am J Trop Med Hyg* 1982, 31: 92–9.

9 T Majima. A strange case of liver cirrhosis caused by parasitic ova. *Tokyo Igakkai Zasshi* 1888, 2: 898–901; F Katsurada. *Schistosomum japonicum*, a new human parasite which gives rise to an endemic disease in different parts of Japan. *J Trop Med Hyg* 1904, 8: 108–11; A Fujinami. Ueber die pathologische Anatomie, und ueber den bom Verfasser entdeckten Weiblichen Parasiten des *Schistosomium japonicum. Kyoto Igakkai Zassi* 1904, i: 1; J Catto. A new blood fluke of man. *Trans Pathol Soc Lond* 1905, 56: 179–89; J Catto. Schistosoma cattoi, a new blood fluke of man. *Br Med J* 1905, i: 11–13.

10 A Fujinami, H Nakamura. The mode of transmission of Katayama disease of Hiroshima Prefecture, Japanese schistosomiasis, the development of its causative worm, and the disease in animals caused by it. *Hiroshima Iji Geppo* 1909, 132: 324–41; K Miyairi, M Suzuki. On the development of *Schistosoma japonicum. Tokyo Iji Shinshi* 1836, 1913: 1–5; R T Leiper. Report on an expedition to China to study the trematode infection of man. *Trop Dis Bull* 1915, 6: 295–6.

11 *Op cit.* See notes 2 and 7 (Low) above.

12 J B Christopherson. Notes on a case of espundia and three cases of kala-azar in the Sudan treated by the intravenous injection of antimoniam tartratum. *J Trop Med Hyg* 1917, 20: 229–36; J B Christopherson. The successful use of antimony in bilharziosis. Administered as intravenous injections of antimonium tartrate (tartar emetic). *Lancet* 1918, ii: 325–7; P Manson-Bahr. In: *History of the School of Tropical Medicine in London.* London, 1956: H K Lewis, pp. 203–4; A Crichton-Harris. Undercurrents on the Nile: the life of Dr John B Christopherson (1868–1955). *J Med Biog* 2006, 14: 8–16. See also *Op cit.* See note 2 above.

13 C. Singer, E A Underwood. *A Short History of Medicine*, 2nd edn. Oxford, 1962: Clarendon Press, p. 493. See also *Op cit.* See note 7 above (Low); H H Scott. Ankylostomiasis. In: *A History of Tropical Medicine.* London, 1939: Arnold, pp. 840–53.

Joseph Everett Dutton (1874–1905): West African trypanosomiasis and relapsing fever

In 1905, Joseph Everett Dutton (Figure 11.1) died at Kasongo, Congo Free State (now the Democratic Republic of Congo) aged 30 years.[1] Despite his short life, he had made an enormous contribution to contemporary understanding of both west African trypanosomiasis and African 'tick fever'.

On 13 September 1903 he had travelled to the Congo in the company of Doctors John Lancelot Todd (1876–1949) and Cuthbert Christy (1864–1932) with the Liverpool School of Tropical Medicine's twelfth expedition. In late 1904 they reached Stanley (now Boyoma) Falls, where Dutton and Todd demonstrated both the cause and mode of transference of 'tick fever' from man to monkeys (Christy had returned to England in June 1904[2]). Both investigators contracted 'tick fever' but were able to proceed with their journey; when they reached Kasongo on 9 February both seemed (according to a letter subsequently received in Liverpool) to be in 'excellent spirits'.[3] However, Dutton's health rapidly deteriorated, and he died on 27 February.

William Osler, speaking in 1909, subsequently coupled his name with those of three other 'true martyrs of [tropical medical] science' – Walter Myers (1872–1901) and Jesse W Lazear (1866–1900), both of whom died of yellow fever (see Chapter 6), and Patrick T Manson (1878–1902). The latter was the elder son of Sir Patrick Manson (see Chapter 3), who died following a shooting accident whilst researching beri-beri on Christmas Island.[4]

FIGURE 11.1   Joseph Everett Dutton (1874–1905) (reproduced courtesy of The Wellcome Library, London).

BIOGRAPHICAL DETAILS AND RESEARCH

Dutton, the twelfth of thirteen children, was educated at the King's School, Chester, and Liverpool University.[5] He graduated MB ChB in 1897, and was appointed to a George Holt Fellowship in Pathology. He was subsequently elected Walter Myers Fellow of the Liverpool School of Tropical Medicine (1901–4). The School's west African expeditions were then in full swing, and were largely financed by Leopold II, King of the Belgians.[6] Dutton's first expedition – he was to undertake four – to southern Nigeria (the School's third) took place in 1900, when he was accompanied by Doctors Annett and Elliott; in 1901 he went alone to the Gold Coast (now Ghana) and The Gambia on the School's sixth expedition, and produced a valuable report on malaria prevention. It was during this trip that he identified *Trypanosoma brucei gambiense* in the blood of a patient under the care of Dr R M Forde (1861–1948; see below), in addition to describing several other trypanosomes.[7] The first physician to describe west African trypanosomiasis was probably Atkins, but credit is usually attributed to Thomas Winterbottom (see Chapter 1). On 21 September 1902, Dutton proceeded (with Todd) to The Gambia and French Senegal, on the School's tenth expedition. His fatal infection was acquired during the twelfth expedition, which began in September 1903.

Dutton was thus the first to describe a trypanosome in the peripheral blood of *Homo sapiens* – in 1901. The infected patient, an Englishman named Mr Kelly

FIGURE 11.2   J E Dutton with Dr R M Forde and the patient (Kelly) from the peripheral blood of whom *T. brucei gambiense* was demonstrated (reproduced courtesy of The Wellcome Library, London).

(see Figure 11.2), was serving as Master of the government steamer on the River Gambia. In May 1901 he had been admitted to Bathurst Hospital with a febrile illness, where Forde – later principal medical officer to the West African Medical Staff – demonstrated 'wriggly worms' (possible of filarial origin) in Kelly's blood. He was not able to identify these organisms accurately, and the patient was therefore sent to Liverpool and examined by Dutton, who was also unable to identify the parasites in serial blood smears. The correct diagnosis therefore eluded them both. During the return voyage to The Gambia the patient's symptoms recurred, and he was examined shortly after arrival (on 15 December 1901), again by Dutton, who had just begun the Liverpool School's sixth expedition. Dutton immediately recognized and described haematogenous *T. brucei gambiense*. Kelly died of cardiac failure on 1 January 1903. During the tenth expedition, Dutton and Todd examined blood from over 1000 'natives'; in only six (all asymptomatic) were trypanosomes found.[8] There was still no suggestion that this organism, subsequently shown to produce 'trypanosoma fever', was also causatively related to the 'negro lethargy' of Uganda.[9]

Dutton died while investigating not trypanosomiasis but (African) relapsing fever, the cause of which (*Spirillum*, now renamed *Borrelia duttoni*) he had also discovered. Tributes subsequently came from many quarters, including the Rt Hon Joseph Chamberlain (Secretary of State for the Colonies), Professor (later Sir) Ronald Ross, Sir Alfred Jones (Chairman of the Liverpool School of Tropical Medicine), and the Administrator of all the French West African Colonies (via Mr Milne, Secretary of the Liverpool School). An obituary notice in *The Times* recorded:[10]

> Death came to him while [he was] working in the hope of learning that which might save hundreds of black men who were dying of sleeping sickness [and] though Dr Dutton [has] passed away, his name [will] always be remembered for the services he [has] rendered mankind.

Correspondence from Dutton to his family in England

The following letters were sent by Dutton (during the Liverpool School's twelfth expedition) from the Congo to his brothers and sisters in England; they were written in September and December 1904, respectively. Although devoted in the main to descriptions of the local countryside, living conditions, etc., both also allude to west African trypanosomiasis:

> *Botali* near Bumba River Congo [Democratic Republic of Congo] (10 miles inland)
> *Sept 1st 1904*
>
> Dear Brothers and Sisters,
>
> It is pitch dark and I am sitting at my camp table underneath the veranda of my tent at 9.30 pm, pen in hand to try to write a letter home – the sentry is passing to & fro in front of me & I occasionally hear the faint click of metal as he shoulders his gun, or perhaps a dry twig snaps as he walks along. Otherwise the night is perfectly still, save the chirp of the crickets & the dull snore of my guard [?] boy asleep behind my tent. I tried to write last mail but the boat (Mail Cargo Steamer) forestalled me by coming two days too soon. Heiberg [Dr Inge Valdemar-Heiberg, a reserve captain in the Norwegian army] & I have taken this little journey into the interior to see if there is any sleeping sickness about, as from our previous work apparently only the river towns are infected.
>
> We arrived here about 12.30 starting from Bumba at 9.30, our lives being so precious the state authorities would not allow us to go without 10 soldiers & a white man, so tonight I shall be guarded when I sleep by a soldier with his gleaming 'bunduk' tipped with steel. The native town we have come to is situated in a broad & extensive bush & grass plain and consist [*sic*] of two rows of long huts separated by a broad pathway & extends [?] for about a mile, there are plenty of people and much of the ground around is taken up with manioc plantations, the women wear only a string of beads round their hips & the men strut about in a loin cloth and tattoo [?] marks on their foreheads. Both rub themselves with curn [*sic*] wood powder & in addition the men have a horrible mixture of buck [*sic*] stuff mixed with oil paint on the necks and forehead. The men were busy making knives at the forge when we entered, the women either scraping manioc or rubbing pieces of curn wood together to get the powder.
>
> Todd has gone with our steamer 'Le Roi des Belges' to a place on the river where buffalo are said to abound, soldiers have also gone with him to see that any native does not kill him!!! He has already shot 2 *beasts* in the morning lower down river, & one harness antelope; in the blood of the latter trypanosomes were present.[11]
>
> We have nearly finished our journey to Stanley pool about a week more on our 'sardine box' as the captain calls our boat & we will start our long journey to Kassongo [*sic*] – I tell you I wont be home for ages – From 'Nouvelle', Anvers [*sic*] up to Bumba we have met with very curious natives, the Bangala and Upoto tribes, whose facial disfigurement (!) is really fine, their faces just resemble the 'coeur de boeuf' so dotted are they all over with small artificial skin prickles (Keloids), the women go about in beautiful ballet [?] skirts of

grass. The belles cut them away behind & dye the lower margin of the skirt red. These tribe [*sic*] were & probably are (on the sly) cannibals.

We have now reached the widest part of the Congo [Zaire] about 20 miles across, but it is impossible to see both banks at one time because of the numerous large islands which divide the river up into numerous channels 1 to 4 miles wide; scenery of Congo here is these [*sic*] wooded islands on the border of which the trees are so covered with creeper that it is often necessary when one wants to land, to cut through a wall of foliage 3 ft thick or more. One fine thing about The Congo is that fish in its waters are plentiful & very excellent eating. There are hosts [?] of different kinds some very curiously shaped with a sort of suckerlike mouth.

Well I must close this letter for tonight.

With love; if I can't write any more before next mail. Goodby.

Jos
Sept 4th

Came back from the 'bush' today, having found two cases of trypanosomiasis amongst the natives of the town of Botsali one a woman who had never left the district the other a man who had often been to fish on the main river where there are plenty of tsetse flys [*sic*] (only one fly was caught in the bush at Botsali).

To Arthur [a brother who was also a medical practitioner] – Our method now of detecting tryps in a person is by puncture of the lymphatic glands. a hypodermic syringe is used, & a drop only of gland fluid obtained. It is easy & apparently causes no pain & is very efficient. See our papers in … and Lancet sent home by last mail. Cervical glands general [*sic*] featured.[12]

Now back on Le Roi des Belges & ½ hours steam brought us to Todds camping ground. He has shot nothing as yet but has found out where the buffalo are & tomorrow he goes with his beater to kill I hope. It is dangerous work.

The mail boat arrived during our absence in the bush & *no letters from home & no papers*. If you wish to write to me address. c/o Monsieur Dohet Leopoldville [now Kinshasa]. After Stanley [Boyonna] Falls we shall be quite out of civilisation & if by chance a letter reaches me it will be very very acceptable.

Now I have my diary to write up, so I must stop for tonight.

Jos

Kasongo
Dec 22nd 04

Dear Brothers & Sisters

I must write this mail; – A happy Xmas to you all. I wish I were at home enjoying Family chat plum pudding & turkey, not forgetting 'that cow again' but still here we wont do badly with our fresh milk & butter & we will have a tinned plum pudding. There is a large herd of cattle here principally remarkable for their huge horns though there is a variety coming from Lac Kivu with short horns & short forequarters so that the back slants upwards from before back; the milk is good but the method of obtaining it is curious; the native cow boy calls his cow 'Liboko' or whatever the name may be (they are all named) ties her by the horns to a post & fastens her hind legs together & merrily starts milking into an old sugar tin: – I don't think I would quite like to see how the butter is made.

The country around here is one huge undulating plain scarcely a tree exists, the land is very rich & would produce anything, potatoes appear to be grown easily & are very good. We have decided to stay here a little time, more especially as we have here a good lab & wish to continue our human tick work, which has turned out exceedingly interesting, we have solved the mystery of the sickness & are naturally delighted.

Since leaving the great forest we have come into the region of wild fowl but have not seen antelope. What will Edward & John say to this, soon after leaving Nyangwe in canoes I came to a sand bank, covered with duck. I shot (choke 12, No 2 shot) & got 9 of them, a little later on another sand bank with geese I got out, crawled behind a clump of grass, shot again (same shot & choke) killed 1 wounded 2 others, got one of the latter easily but the other lost owing to foolishness of boys paddling who with excitement, stripped naked & all made a bold for dead goose instead of watching the one wounded in the river; these geese weighed 14 lbs each. They are black & white with red heads.

Our boys have been made happy during the past 4 weeks with 3 hippos & 1 buffalo; it is a sight to see the natives attacking & hacking up a hippo, they think the flesh a great treat.

You will be astonished when I tell you though we are in the centre of Africa we are within 6 days communication with England. From here there is a telephone line to Baraka on the Gulf of Burbon [*sic*] [erasure by JED] Lake, Tanganyika, from there a canoe takes the message sent by telegraph around the head of the Lake to a port on the German side, & from here it goes by wire to Zanzibar & so home. We have sent a message to the [Liverpool] School this way, the time is taken up crossing the lake. What strides the State has made:- only 12 years ago Baron Dahnis [?][13] was waging war against Sefu, the son of Gipo Gib[14] & other big Arab slave raiders. Kasongo was the principal base during the latter part of the war. A Dr Hinds [*sic*], an Englishman wrote a book about the Fall of the Congo Arabs he was seeing at the time under Dahnis [*sic*] [S Hinde, *The Fall of the Congo Arabs, 1897*], it is worth reading to those interested.

To Ellie [JED's sister]. Very glad to receive your letter, please congratulate Linley and Margie for me I was delighted to hear of their success.

To Tom [JED's brother]. Thank you for your note though short. Have you received stamps?

To all the others except Hugh [another brother of JED]. Received no letters for ages.

Love to all
Your affectionate brother
Jos[15]

Dutton was destined to die in February 1905, probably as a result of tick-borne relapsing fever (see above). According to the obituary notices, Dutton persisted in working during his convalescence from a severe attack of spirillum fever, which he was at the time investigating.[16] The final 'attack' was sudden and he was unconscious for four days before he died, Drs Todd and Heiberg being present.[17]

AN ACCOUNT OF DUTTON'S DEATH

The day after his death, Todd sent a typewritten letter to Dutton's brother, H E Dutton, from Kasongo, conveying this record of Dutton's death:[18]

Kassongo [*sic*]
Congo Free State
Feb 28th 05

Dear Mr Dutton,

I write to you because you are the only one of my poor friend's relatives whose address I know. Please consider I address all of that family who loved him. You have long since been told – that your brother, my partner, is dead. Dr Heiberg and I, his – comrades, have written a long account of his illness to Dr Annett & have asked him to tell you how poor 'Dutt' died.

'Relapsing fever' had left him weak, though he would not admit it & insisted on working as hard as usual. When the breakdown came, he had not the strength to live through it. It is but scant comfort, still, it is well to know that he certainly suffered but little; he was so long unconscious, and that his last certain moment of consciousness was pleasant; he laughed a good morning to Heiberg on the twenty third.

During, and after, his illness the officers of this port did all in their power to serve [?] him. We did what we could, but he became – quickly worse & died, tired out, last night at nine o'clock.

To-day is very quiet, all work is stopped on the Station. There is no noise. The flag is flying low & all the dozen white men have come to say goodbye to Dutt & to tell us how deeply they regret his loss.

There are no clergymen here; seven miles away there are Roman Catholic priests, 'White Fathers'. Dutt has frequently told me that his father has been a strict churchman, so we thought it best that I, the only – English speaking European here, should try to read part of the Church of England Burial Service over him. In the afternoon, after the heat of the day, the soldiers carried him to the cemetery & we buried him.

He lies in the centre of the Africa that holds his life's work.

We have taken photographs of him, his grave, & the surroundings in which he passed his last months. Some of the plates we shall develop ourselves, others we shall send to W H Tomkinson, Dale St. to be developed. I have asked him to send you as many prints as you may wish. I shall bring you home his guns, his bicycle, his instruments and books, his trinkets and a part of his clothing. I shall give the remainder of his clothing and his bedding to the different boys who served him, as keepsakes from their master.

The grief of his boy Ooma was heart-breaking. I shall never forget his sobbing calls to the master who could no longer hear him. Soon after Dutton engaged the lad, he became ill. He had large abscesses all over his body, and for several months the boy was unable to leave his bed. Dutt looked after him, treated him, in fact saved his life half a dozen times over. I think that Ooma was grateful.

Your brother and I had planned work sufficient to keep us in Africa until the middle of next July. I'm going to finish that, our common work, before returning to Europe. So soon as I do – arrive in Liverpool, I shall do him a last service in bringing to you his curios and belongings. It is useless to say that I grieve for him. A week ago, I was rich in a friend, a comrade of over two years work. Now – I am alone.

I am, Yours sincerely,

J L Todd

Obituary notices indicate that there were 12 pall bearers (all 'native sergeants'), and over 1000 individuals turned out to see Dutton's coffin pass to its final resting place; a farewell eulogy was delivered (in French) at the graveside by Commandant Verdick.[19]

Figure 11.3 shows the sites at which Dutton demonstrated the aetiological agents of West African trypanosomiasis and African 'tick fever'.

A lasting memorial

The Dutton family (of Cheshire) subsequently placed a memorial window in Bunbury Church; however, this was destroyed by enemy action in November 1940. The original window had been dedicated on 17 January 1906 by the Dean of Chester, Dutton's father and many of the Liverpool School's staff being

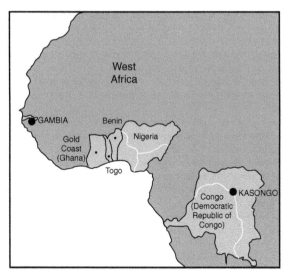

FIGURE 11.3 Map showing the sites of Dutton's major researches.

present. The replacement window was dedicated by the Archdeacon of Chester on 7 December 1952.

NOTES

1 Anonymous. Joseph Everett Dutton. *Lancet* 1905, i: 1239–40; Anonymous. Joseph Everett Dutton. *Br Med J* 1905, i: 1020–21, 1072, 1250, 1363; Anonymous. Dr J E Dutton. *Times* 1905, 1 May: 6; 19 May: 10; H H Scott. *A History of Tropical Medicine*. London, 1939: Arnold, pp. 1027–8; J Braybrooke, G C Cook. Joseph Everett Dutton (1874–1905): pioneer in eluci-dating the aetiology of west African trypanosomiasis. *J Med Biog* 1997; 5: 131–6; P J Miller. Applying the lessons overseas. In: *'Malaria, Liverpool': an illustrated history of the Liverpool School of Tropical Medicine 1898–1998*. Liverpool, 1998: Liverpool School of Tropical Medicine, pp. 23–6; D'A Power, M Bevan. Dutton, Joseph Everett (1874–1905). In: H C G Matthew, B Harrison (eds), *Oxford Dictionary of National Biography*, Vol. 17. Oxford, 2004: Oxford University Press, pp. 443–4.

2 *Ibid.*

3 *Ibid.*

4 G C Cook. William Osler's fascination with diseases of warm climates. *J Med Biog* 1995, 3: 20–9.

5 *Op cit.* See note 1 above.

6 *Liverpool School of Tropical Medicine: Historical Record 1898–1920*. Liverpool, 1920: Liverpool University Press, p. 103. See also A Hochschild. *King Leopold's Ghost: a story of greed, terror, and heroism in Colonial Africa*. London, 1998: MacMillan, p. 366.

7 *Op cit.* See note 1 above. See also: T Winterbottom. *An account of the native Africans in the neigh-bourhood of Sierra Leone, to which is added an Account of the Present State of Medicine Among Them*, 2nd edn, two volumes. London, 1969: Frank Cass, pp. 362, 383; J C Dutton. Preliminary note upon a trypanoma occurring in the blood of man. *Thompson Yates Laboratories Report* 1902, 4: 455–70; G C Low. A retrospect of Tropical Medicine from 1894 to 1914. *Trans R Soc Trop Med Hyg* 1929, 23: 213–32.

8 B I Williams. African trypanosomiasis. In: F E G Cox (ed.), *The Wellcome Trust Illustrated History of Tropical Diseases*. London, 1996: Wellcome Trust, pp. 178–91; F E Cox. History of sleeping sickness (African trypanasomiasis). *Infect Dis Clin North Am* 2004, 18: 231–45. See also *Op cit*. See note 7 above.

9 G C Cook. Correspondence from Dr George Carmichael Low to Dr Patrick Manson during the first Ugandan sleeping sickness expedition. *J Med Biog* 1993, i: 215–29; G C Cook. Sir David Bruce's research on trypanosomes. *J Med Biog* 1996, 4: 61; B I Williams. African trypanosomiasis. In: F E G Cox (ed.), *The Wellcome Trust Illustrated History of Tropical Diseases*. London, 1996: Wellcome Trust, pp. 178–91.

10 *Op cit*. See note 1 above. See also: H H Scott. Relapsing fever. In: *A History of Tropical Medicine*. London, 1939: Arnold, pp. 781–94.

11 This demonstration of trypanosomes in an animal reservoir is of interest; today only limited evidence exists (in pigs and dogs) for a zoonotic reservoir of *Trypanosoma brucei gambiense*.

12 *Op cit*. See note 7 above (Winterbottom), pp. 29–31. This is an early description of the use of this technique in the demonstration of *T. brucei gambiense*. It remains in use today, enlarged cervical glands (first described by Winterbottom in 1803) being those most frequently aspirated.

13 JED presumably refers to Baron Dhanis, whose anti-slavery campaign was at this time centred on Kasongo. T Pakenham. *The Scramble for Africa 1876–1912*. London, 1991: Weidenfeld and Nicolson, pp. 434–51.

14 JED probably means Tippu Tip, whose son Sefu was one of the last of the great Arab slave traders – whose activities were centred on Zanzibar; *Ibid.*, pp. 434–5.

15 These two letters are in the possession of Josephine Braybrooke, a niece of J E Dutton, at Woodbridge, Suffolk. See note 1 above (Braybrooke, Cook).

16 *Op cit*. See note 1 above.

17 The mode of transmission of relapsing fever in Dutton's case remains unclear; it seems highly likely, however, that he contracted the infection while carrying out a post-mortem examination on an infected African patient. African tick-borne relapsing fever is caused by *Borrelia* spp., and transmitted by soft ticks (genus *Ornithodoros*). Although the mortality rate in untreated infection is low compared with that for the louse-borne variety, being less than 10 per cent, neurological complications are relatively common in the non-immune although rare in the indigenous population. The infection is now amenable to tetracycline or penicillin chemotherapy; a Jarisch–Herxheimer reaction is a rare complication.

18 J L Todd to H E Dutton (brother of JED) 28 February 1905. Manuscript collection of Josephine Braybrooke.

19 *Op cit*. See note 1 above.

12

The causative agent of visceral leishmaniasis (kal-azar): William Leishman (1865–1926) and Charles Donovan (1863–1951)

Although their names have been linked by a hyphen for a hundred years or thereabouts, Leishman and Donovan were very different individuals from different backgrounds; furthermore, when the Leishman–Donovan body (now the amastigote) of kal-azar was discovered they were working on different continents.

The causative agent of kala-azar was first demonstrated in 1903; in the same year, the organism causing Delhi Boil (which has various other local names), *Leishmania tropica*, was described by James Wright (1870–1928). In fact, the organism had been described five years earlier by Petr Borovsky (1863–1932), but unfortunately his article was written in Russian and hence unknown in the West.

WILLIAM BOOG LEISHMAN (1865–1926)

William Boog Leishman Kt, FRCP, FRS, KCMG (Figure 12.1)[1] was a bacteriologist and pathologist who was born in Glasgow on 6 November 1865. His father (also William; 1833–94) was Regius Professor of Midwifery in the University of Glasgow; an uncle was Thomas Leishman (1825–1904), a Church of Scotland minister. Leishman was educated at Westminster School and the University of Glasgow, from which he graduated MB, CM in 1886. The following year he

FIGURE 12.1 Lieutenant-General Sir William Boog Leishman (1865–1926) (reproduced courtesy of The Wellcome Library, London).

entered the Army Medical Service and passed into the Army Medical School at Netley. He spent three years in England, after which he was posted to India; apart from a one-year period of sick leave (1892–3) he spent seven years there during which he served in the Waziristan campaign (1894–5).

In 1899 Leishman was promoted to the rank of Major and posted back to Netley, where he had charge of the medical wards. Whilst there he also worked in the laboratory supervised by Sir Almroth Wright, Professor of Pathology. In 1900, he succeeded Wright in this Chair.

In 1900 Leishman demonstrated the causative agent of kala-azar (dum-dum fever), but he did not publish this finding until 1903. In that year (1903), Donovan of the IMS (see below) confirmed this discovery at Calcutta. Ross (see Chapter 5) later termed the disease caused by Leishman–Donovan bodies 'the leishmaniases'.

Leishmaniasis was not, however, the only disease in which he made a significant contribution. In 1901 he simplified the diagnosis of malaria by introducing Romanowsky staining, and the following year he developed a method for quantifying the phagocytic properties of blood. In 1903 the Army Medical School was transferred from Netley to Millbank, where until 1914 he occupied the Chair of Pathology; he continued his work on leishmaniasis and also worked on a typhoid vaccine (which was administered to British troops in the First World War). Following this war, he spent some time elucidating the life-cycle of *Spirochaeta duttoni* (see Chapter 11) – the causative agent of relapsing fever.

In January 1914 Leishman became the War Office's expert on tropical diseases, and in October of that year he joined the British Expeditionary Force and organized the pathological services in France and Flanders. He also chaired War Office committees on 'trench fever' and 'trench nephritis'. Returning to practice in England in April 1918, he was soon gazetted Major-General; in June 1919 he became the first director of pathology at the War Office, and in July 1923 he was appointed Director-General of the Army Medical Services and advanced to the rank of Lieutenant-General.

In 1912 Leishman became honorary physician to King George V. He received numerous honours, both in Britain and abroad. Leishman was a member of the Medical Research Committee (later Council) and third President of the (Royal) Society of Tropical Medicine and Hygiene (1911–12). His Presidential Address to the newly-formed Society dwelt entirely on the role of the Society in the future, stating that it should he maintained as a centre located in London, which should be a meeting point for Fellows worldwide with a tropical interest. Leishman died at the Queen Alexandra's Military Hospital, Millbank, in June 1926 following a short illness, and was buried in Highgate cemetery.

CHARLES DONOVAN (1863–1951)

Charles Donovan MD (Figure 12.2)[2] was an Irishman, who graduated from Trinity College, Dublin, with an MD in 1889. He entered the IMS two years later. He served on the North-West Frontier with the Tirah Expeditionary Force, and subsequently took part in the action at Dargai and operations in the Bara Valley. He was then posted to Madras for civil employment, and shortly afterwards was appointed to the Chair of Physiology at the Madras (now Chennai) Medical College – which he held until his retirement. He was also appointed second (and subsequently senior) physician at the General Hospital, and Superintendent of the 80-bedded Royapettah Hospital. His discovery of the causative agent of visceral leishmaniasis (see below) was made before organized, or indeed any medical research existed in India.

Donovan is said to have possessed a dynamic and original personality, and was an excellent clinical teacher and microscopist. In addition, he had a well-developed sense of humour, ready sense of wit, and encouraged inquiry in his

FIGURE 12.2 Charles Donovan (1863–1951) (reproduced courtesy of The Wellcome Library, London).

students. Professor A S Mannadi Nayar remembered Donovan as 'an inspiring professor, an inquiring physician, an original teacher, [and] a great scientist'.

Donovan retired from the IMS in 1920 with the rank of Lieutenant-Colonel, and thereafter concentrated on his hobbies; he was an expert on the lepidoptera, and was also a talented amateur artist. He lived to the ripe age of 88 years.

THE CAUSATIVE AGENT OF KALA-AZAR[3]

Prior to 1823 there were few accurate records of visceral leishmaniasis; however, in the following year an outbreak of a chronic fever associated with marked cachexia and anaemia, which was not cured with quinine, occurred at Jessore in Bengal. This illness – Bardwan fever, kala-azar or 'black disease' (derived from the dark discolouration) – reached Bardwan in 1862. In 1869, it appeared in the Garo Hills, Assam. In 1889, Giles considered that the anaemia was caused by hookworm (a contested view) and beri-beri, and that the splenomegaly was malarial. Rogers (see Chapter 13) was sent to Assam in 1896 to investigate the disease,[4] and two years later Ross (see Chapter 5) followed; both concluded that it was caused by a virulent form of malaria (a view which was upheld for several years), despite the fact that there was little malarial pigment and few, if any, malarial parasites.

In 1900 at Netley, Leishman used his modified Romanowsky technique to identify protozoan parasites, which he incorrectly identified as trypanosomes, in the spleen of a soldier from Dum Dum, Calcutta. Various protozoologists had differing views on the identity of these organisms, and Leishman did not therefore publish his findings until 1903.[5]

In July 1903, Donovan sent a sketch of the parasites he had found in the spleen of a 'kala-azar' patient to Ross. It became clear that both Leishman and Donovan had visualized identical organisms; Ross had originally created the genus *Leishmania* (in honour of the discoverer) in 1903; however, he later claimed that the investigators had made the discovery independently, and from then onwards they were designated Leishman–Donovan bodies.[6]

By 1904, it had become clear that these protozoan parasites were identical with those described by Wright (see above) in oriental sores in 1903.

TRANSMISSION

Rogers (see Chapter 13) cultured the parasite and, finding the flagellate stage, suggested that an extra-corporeal phase might exist – possibly in the bed-bug. Although the sandfly (*Phlebotomus* sp.) soon came under suspicion, it was not until 1942 that this vector was definitively confirmed as conveying kala-azar.[7] As early as 1911, however, Charles Wenyon (1878–1948) had suggested that this constituted the transmitter of cutaneous leishmaniasis.[8]

NOTES

1   Anonymous. Sir William Leishman: inoculation against typhoid. *Times, Lond* 1926, 3 June; Anonymous. Sir William Boog Leishman. *Lancet* 1926, i: 1171–3; Anonymous. Lieut-General Sir William Leishman. *Br Med J* 1926, i: 1013–16; Anonymous. Army's chief 'M.O.' dead: Sir W Leishman's war on microbes. *Evening News, Lond* 1926, 2 June; H H Scott. William Boog Leishman (1865–1926). In: H H Scott (ed.), *A History of Tropical Medicine*. London, 1939: Arnold, pp. 1058–62; Anonymous. Leishman, Sir William Boog. *Munk's Roll*, Vol. 4. London, Royal College of Physicians, pp. 538–9; H D Rolleston, H J Power. Leishman, Sir William Boog (1865–1926). In: H C G Matthew, B Harrison (eds), *Oxford Dictionary of National Biography*, Vol. 33. Oxford, 2004: Oxford University Press, pp. 283–5.

2   R Christophers. Charles Donovan. *Br Med J* 1951, ii: 1158; Anonymous. Charles Donovan. *Br Med J* 1951, ii: 1286.

3   G C Low. A retrospect of Tropical Medicine from 1894 to 1914. *Trans R Soc Trop Med Hyg* 1929, 23: 213–32; C Singer, E A Underwood. *A Short History of Medicine*, 2nd edn. Oxford, 1962: Clarendon Press; G C Cook. Some early British contributions to tropical disease. *J Infect* 1993, 27: 325–33. P E C Manson-Bahr. Old World leishmaniasis. In: F E G Cox (ed.), *The Wellcome Trust Illustrated History of Tropical Diseases*. London, 1996: Wellcome Trust, pp. 206–17.

4 L Rogers. *Report of an Epidemic of Malarial Fever in Assam, or Kala-azar*. Shillong, 1897: Assam Secretariat Printing Office, p. 223. See also A Castellani, A J Chalmers. *Manual of Tropical Medicine*. London, 1919: Baillière, Tindall and Cox, p. 2436.
5 W B Leishman. On the possibility of the occurrence of trypanosomiasis in India. *Br Med J* 1903, i: 1252–4; G C Low. A retrospect of Tropical Medicine from 1894 to 1914. *Trans R Soc Trop Med Hyg* 1929, 23: 213–32.
6 C Donovan. On the possibility of the occurrence of trypanosomiasis in India. *Br Med J* 1903, ii: 79; R Ross. Notes on the bodies recently described by Leishman and Donovan. *Br Med J* 1903, ii: 1261–2.
7 C S Swamminath, H E Shortt, L A P Anderson. Transmission of kala-azar to man by the bite of *Phlebotomus argentipus*. *Indian J Med Res* 1942, 30: 473–7.
8 C Singer, E A Underwood. *A Short History of Medicine*, 2nd edn. Oxford, 1962: Clarendon Press, pp. 487–8.

13

Leonard Rogers (1868–1962): the diseases of Bengal, and the founding of the Calcutta School of Tropical Medicine

Sir Leonard Rogers FRS (1868–1962; Figure 13.1) was one of the most significant pioneers of tropical medicine. Born at Helston, Cornwall, and educated at Tavistock Grammar School, Plymouth College, and St Mary's Hospital, London, he joined the Indian Medical Service (IMS) in 1893, primarily as a pathologist. Most of his research was carried out in eastern India, to which he travelled in 1893 at the age of 26 years. His expressed intention was to devote himself 'to research in tropical diseases' – which he accomplished over the following fourteen years.[1] The disease spectrum in Bengal had already been delineated by William Twining (1790–1835), amongst others, in the early nineteenth century (see Chapter 1), but the causes of the various entities were still not understood.[2] His important book on *Fevers in the Tropics* was published in 1908.[3] Whilst on leave in England that year he was offered the Chair of Tropical Medicine at the Royal Army Medical College; however, he chose to return to India and continue research, much of which focused on cholera and amoebic dysentery. It was during these years that he was also the major figure leading to the foundation of the Calcutta School of Tropical Medicine, which caused him a decade of problems – mostly related to fund-raising. A young doctor, McCabe-Dallas, working in Assam was responsible for the original idea for this project – a fact which most authors have either ignored or understated.

FIGURE 13.1 Sir Leonard Rogers (1868–1962): photograph by J Russell & Sons (reproduced courtesy of The Wellcome Library, London).

Bengal

Rogers initially investigated kala-azar (visceral leishmaniasis) in Assam, where this disease was endemic, and came to the conclusion that it was a form of chronic malaria. He also used antimony in kala-azar later, with moderate success.[4] In 1900, Rogers was appointed Professor of Pathology in Calcutta (Kolkata); here he worked on *Entamoeba histolytica*, which he correctly associated with both dysentery and hepatic 'abscess', and pioneered emetine (a component of ipecachuana) treatment for this infection – *in vitro* work in the USA had already indicated that this compound was toxic to *E. histolytica*.

E. histolytica had been first described by Friedrich Lösch (1840–1903) in 1875 in the faeces of a Russian peasant from Archangel, and liver 'abscess' had been described by William Budd (1811–80) in 1845. C M Wenyon (1878–1948) and Clifford Dobell (1886–1949) were later responsible for a great deal of work on *E. histolytica* during the Great War (1914–18).[5]

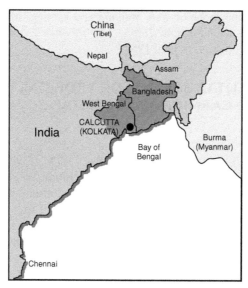

FIGURE 13.2   Map showing Bengal and Calcutta (Kolkata) – the site of Rogers' School of Tropical Medicine and Hygiene.

In cholera, which was endemic in India, Rogers researched the best composition of fluid for intravenous rehydration.[6] He also assessed the value of sodium gynocardate (a component of chaulmoogra oil) in the treatment of leprosy. Other contributions were research on snake venoms, and on trypanosomiasis in horses. Above all, he took a great interest in the epidemiology of all of the major infections of India.[7]

Figure 13.2 shows the position of Bengal and the site of Rogers' School of Tropical Medicine and Hygiene, in Calcutta (Kolkata).

Life in London

Following retirement from the IMS in 1920, still aged only 52 years, Rogers was appointed Medical Adviser to the Secretary of State for India in London. He was also elected extra physician to the Hospital for Tropical Diseases, then situated at Endsleigh Gardens, WC1, and in addition was a lecturer at the London School of Tropical Medicine. Rogers also devoted much time to raising money (he was now an experienced fund-raiser) for the British Empire Leprosy Relief Association. Rogers was knighted in 1914 and elected to the Royal Society in 1916. From 1933 to 1935 he was President of the Royal Society of Tropical Medicine and Hygiene. At the Royal College of Physicians he gave the Milroy (1907) and Croonian (1924) lectures, and he received the Moxon medal in 1924.

Rogers' principal hobbies were walking and cycling. Much of his life is encapsulated in his autobiography *Happy Toil: fifty-five years of tropical medicine*, which was not published until 1950.[8]

THE CALCUTTA SCHOOL OF TROPICAL MEDICINE AND THE CARMICHAEL HOSPITAL (see Figure 13.3)

A new initiative

On 10 March 1910, Dr Alfred McCabe-Dallas,[9] DTM (Liverpool), an Assam tea-plantation medical practitioner, wrote to the Editor of *The Englishman*, a daily Calcutta newspaper. This letter, which was published the following day, called for a Tropical School which would be of benefit to Assam and Eastern Bengal, and also the 'whole of India'. It should, he felt, be situated in Assam, which 'could be easily approached from all parts [of India] by railway'. It is, he wrote, 'an undoubted fact that a large percentage of mortality in India generally, and Assam [where the disease spectrum was already well known] especially, is a preventable one ...'. He continued, 'It seems an anomaly for medical men to have to go to London or Liverpool to study tropical disease ([where the] clinical material is dependent on the shipping from the East, Africa, and the West Indies), but there is no other option under present circumstances'. He proceeded to emphasize that in those two English Schools it was the laboratory (rather than clinical) work which was of paramount importance. 'Australia, Germany, and the Grand Canary islands have [he wrote] already got their tropical schools, and New Orleans is to have one shortly.' Staffing of his proposed School could

FIGURE 13.3   The School of Tropical Medicine and Hygiene Institute, Calcutta. (Anonymous. *A Short Account of the Calcutta School of Tropical Medicine, Institute of Hygiene and the Carmichael Hospital for Tropical Diseases* 1922, p. 1.)

be 'safely left in the hands of ... Rogers [see above], and Christophers[10] [and] Stephens[11] of Liverpool'. The Indian Government, he claimed, contributed nothing to the two [English] schools and he implied that they would also be unlikely to make a pecuniary contribution to his envisaged school; instead, 'Princes, native noblemen, rich semindars, and wealthy merchants' were more likely to finance an 'Indian tropical school', which could grant diplomas as was apparently done at 'the Liverpool and Cambridge Universities' and also accept 'a thesis in tropical medicine for the M.D Degree'.[12]

Rogers accepts the challenge

This letter from McCabe-Dallas prompted Rogers to write an editorial for *The Englishman*, which was published on 21 March 1910. In it, he proved far more supportive of the Indian Government than was McCabe-Dallas, and gave them 'credit ... for what they [had] done to facilitate and assist the researches of recent workers'. He also maintained that McCabe-Dallas had 'overlooked the fact that in the Medical College, Calcutta [there already existed] laboratories that are the admiration of visitors from every part of the world ...'. Important research had also been carried out in Bombay, Madras and Kasauli; the problem was that at all of these locations the laboratory was far distant from the hospital. He agreed that 'there is [an] urgent need for a fully equipped and fully recognized School of Tropical Medicine for India', but, he felt that 'It should be located in a hospital in one of the great centres so that investigators should be in intimate touch with patients'. He [naturally] favoured Calcutta, where, he felt, 'All that is required is a small extension of staff and a small extension of laboratory accommodation ...'. The trained doctor, he also considered, should be able to diagnose and manage not only local diseases but also 'foreign' tropical conditions such as African trypanosomiasis. He concluded his editorial: 'We are confident that the Government of India will take the matter up in earnest and will carry out the minor alterations and additions necessary to convert the [Calcutta] Medical College into a first class school for research and post-graduate study in tropical medicine'.[13]

By 1910, the London (see Chapter 3) and Liverpool (see Chapter 20) Schools of Tropical Medicine had been in existence for a little more than a decade, and similar institutions were springing up in all those countries with colonial interests. In Calcutta (which was, between 1772 and 1912, the administrative capital of India) there was a large number of medically qualified Indians and expatriates who were not part of the IMS, and therefore a government-funded School would, in Rogers' estimation, make good sense.

Although Rogers acknowledged the primary impetus to his project in his autobiography, he avoided mentioning McCabe-Dallas by name:

> in March 1910, I read a letter in a Calcutta paper, from an Assam plantation doctor, advising the formation of a research institute in that province to investigate diseases affecting the tea-garden labour forces.

This editorial was followed by one in the *British Medical Journal* in April 1910, which he also wrote anonymously, and which gave further credence to the idea: Calcutta, he contended, already had almost 1200 beds and the Medical College Hospital (MCH) there possessed 'the premier hospital and medical school in the East'. The suggestion for a tropical school principally to carry out research into local diseases, as put forward by a 'plantar's doctor in Assam', was to be welcomed, for the MCH already possessed 'suitable laboratory accommodation'. He called on the 'Government of India to found an efficient school of tropical medicine'; this would be a valuable addition to the 'English schools of tropical medicine [which] have done invaluable pioneer[ing] work, and are essential for the training of those about to enter on their labours in hot countries'.[14]

And he was rightly of the opinion that neither of the English schools at the Albert Dock Hospital (ADH)[15] and Liverpool[16] had anything like the concentration of 'tropical' cases[17] as did Calcutta.

Funding of any project is at present full of problems, but the difficulty experienced by Rogers in obtaining financial backing for what might be considered to be a major priority for the Raj (i.e. elucidation of the causes and satisfactory treatment of the multiplicity of disease entities in Bengal which afflicted the 'coolies' and the expatriate alike) now seems impossible to comprehend.

Frustrated attempts

King Edward VII (1841–1910) died on 6 May 1910. Rogers immediately made a proposal that a School of Tropical Medicine in Calcutta, together with a sanatorium, would constitute an appropriate memorial; this idea was received enthusiastically by the Medical Section of the Asiatic Society of Bengal. The idea was also warmly welcomed by Sir Pardey Lukis,[18] then Director General of the Indian Medical Service (IMS) and based at Simla; the Lieutenant Governor (Sir Edward Baker)[19] apparently favoured a non-medical proposal (which was in fact rejected, largely because it failed to find favour with the public, as did Rogers' renewed attempt).[20]

The Government of India felt that rather than launching into a major scheme, which would be very costly, it should begin on a small scale. The Professors of Medicine and Pathology at the MCH should, they considered, institute a six-month course for a Diploma in Tropical Medicine. In January 1911, Lukis told Rogers, the principal of the Medical College and also the Professor of Biology, that there was a possibility that the Government of India might make a grant of 3 lakhs (£21 300) for the project (although in the event this grant did not materialize). The three of them formed a planning committee, but their proposal for a 260-foot frontage was rejected by the Government of Bengal, which would not sanction a site more than 100 ft long.

Further, progress had to await until late 1912, when Lukis persuaded the Government of India to make a grant of 5 lakhs (£37 500) on condition that the Government of Bengal supported Rogers' major scheme (i.e. a 260-ft frontage), which it agreed in principle to do. The site was cleared and Rogers launched an 8 lakhs (£59 000) appeal in September 1913. The Bengal Government architect, H A Crouch,[21] had already drawn up plans. There was then a substantial obstruction to the progress of Rogers' scheme in that an 'Improvement Trust' decided that, in order to widen the main road through Calcutta, it required a portion of the land which Rogers had earmarked for the laboratory. Nevertheless, building of the laboratories began in January 1914, and on 24 February Lord Carmichael,[22] the Governor of Bengal, laid the foundation stone. However, more 'political' problems with the Improvement Trust were to ensue, and in these Rogers was supported by Edwards[23] (Head of the Bengal Medical Department) and Duke[24] (the Indian Civil Service Member of Council in charge of the Medical Department). Crouch had subsequently to redesign the laboratory building.

The next problem involved staffing, and Rogers' request for six professors was backed by Lukis but, unfortunately, not by the Government of Bengal.

The staffing dispute

In April 1914, the issue of staffing Rogers' School of Tropical Medicine came to the fore. The Government of India was not prepared to pay for staff, in addition to their capital contributions to the building, and the Bengal Government also felt unable to finance them. *The Englishman*, on the fourth of that month, contained an anonymous article with a somewhat controversial heading (see Figure 13.4). 'The Government [of India, it claimed] leans to the opinion that the College should be staffed largely, if not exclusively, by the IMS [already in post]' instead of Rogers' plan, which was to 'appoint research officers from amongst men of note and established reputation in the world of medicine'. This might mean, therefore, the article continued, that Rogers' intention to establish a School of Tropical Medicine in Calcutta might have to be abandoned for he would not approve of anything but the best! In fact, Rogers wished to appoint the most suitable individuals worldwide to the proposed new full-time Chairs. This scenario was all the more disappointing because a mere two months previously the Foundation Stone had been laid by the Governor of Bengal (see above).[25]

Three days later, *The Englishman* published a '[Bengal] Government Communiqué' totally denying the accuracy of the article of 4 April; any delay in the opening the School would not be the result of a staffing crisis but, it claimed, would be due to the fact that 'in one part of the site of the proposed school considerable difficulties have had to be overcome in obtaining a safe foundation'. Perhaps not surprisingly, this gave rise to a leading article the same day (7 April). In this, the Government Communiqué was designated 'one of the most misleading official documents that any provincial Government has yet had the audacity to publish'. The article referred to a speech by Colonel Harris at the Foundation Stone ceremony, in which he stated that 'a whole time staff of

TROPICAL SCHOOL OF MEDICINE.

Dispute Over Staffing.

SENSATIONAL DEVELOPMENT.

POSSIBLE RESIGNATION OF COLONEL ROGERS.

The Englishman understands that a serious difficulty has arisen in connection with the proposed School of Tropical Medicine in Calcutta, which may make it impossible to realise the scheme on the plan drawn up by Lieutenant-Colonel Leonard Rogers, I.M.S., if, indeed, on any plan at all commensurate with the expectations that have been raised.

FIGURE 13.4 Heading of an anonymous article in the Calcutta daily newspaper *The Englishman*, published on 4 April 1914.

not less than six Professors will be required …'. Part at least of the problem, the article claimed, was that Calcutta had recently ceased to be the 'Imperial Capital', and the Government of India was putting more planning and finance into Bombay, where it had already sanctioned a new School of Tropical Medicine. Therefore, the Government of India 'is attempting to throw the financial responsibility it morally owes to the Calcutta School …'. People in Bengal, it seems, had supported Rogers' proposed school financially; *The Englishman* felt that 'The Government must do its duty too and help towards the complete realization of the scheme it had originally in view'.[26] Rogers himself therefore had to launch, with the Bengal Government's blessing, another Public Appeal for funds – in the event, highly successful – and he therefore became Honorary Secretary of The School of Tropical Medicine Endowment Fund.

FIGURE 13.5   The Carmichael Hospital for Tropical Diseases, Calcutta. (Anonymous. *A Short account of the Calcutta School of Tropical Medicine, Institute of Hygiene and the Carmichael Hospital for Tropical Diseases* 1922, p. 9.)

The Carmichael Hospital (see Figure 13.5)

Although the Surgeon-General and the Government of Bengal approved the idea of a Hospital for Tropical Diseases, finance was still not forthcoming from official sources and Rogers' Endowment Fund had to raise 2.5 lakhs of rupees. Several major contributions were received, and the Calcutta millionaire, Sir David Yule,[27] proved extremely helpful. Carmichael laid the Foundation Stone of the new hospital, which was named after him, on 24 February 1916 – two years after that of the laboratories.

An Institute of Hygiene

Both Rogers and the Bengal Government decided that the new Calcutta School should teach Hygiene (Public Health) in addition to Tropical Medicine. Again, Lukis was an influential supporter. In the event, it was decided to initiate an Institute of Hygiene to mark the retirement of Carmichael from the Governorship of Bengal; he was succeeded by Lord Ronaldshay (later the Marquis of Zetland). However, this scheme had to be abandoned, largely due to the onset of the Great War (1914–18). After a long struggle to raise funds (in which the new Governor proved helpful) and to remove the police morgue from the site, the Hygiene extension was at last begun in mid-1918.

Completion at last

Figure 13.6 shows the completed school (left) and Carmichael Hospital soon after they were built. The whole project had been completed by 1920.[28] Rogers, who had fought so hard and long to see the project completed, and whose health by this time had deteriorated, was soon to set sail for England, and he did not return to Calcutta.

However, research was now well established into many of the local diseases of Bengal, including kala-azar, malaria, dysentery, leprosy, hookworm disease and diabetes. Subsequent annual reports indicate the valuable results provided by these varied research projects.[29]

Power has concluded that, despite his invaluable researches into many of the diseases of Bengal, 'the Calcutta school [which was in fact, the "brain-child" of McCabe-Dallas, remains] the most tangible legacy of his career'.[30] However, had it not been for the initiative of the young plantation practitioner and the editorial staff of *The Englishman*, this project might never have got under way. This once again emphasizes that the catalytic input of a young, and relatively unknown worker is easily obscured by a more senior and well-established figure.

Rogers thus made substantial advances in 'medicine in the tropics' (he carried out important research into the diseases of Bengal and also established what was probably the first School of Tropical Medicine in a developing country) and also to the formal discipline following his return to London in 1920.

THE CARMICHAEL HOSPITAL FOR TROPICAL DISEASES.
(THE SCHOOL FRONTAGE ON THE LEFT).

FIGURE 13.6 The School and Hospital, Calcutta. (*Annual Report of the Calcutta School of Tropical Medicine Institute of Hygiene and the Carmichael Hospital for Tropical Diseases for the year 1923* 1924, p. 5.)

ROGERS' PRESIDENTIAL ADDRESS TO THE ROYAL SOCIETY OF TROPICAL MEDICINE AND HYGIENE

Rogers began by informing his audience that he started 'investigating the relationship of climatic conditions to the seasonal incidence of disease in India' in 1895. He had found 'a close relationship between the rise of the ground-water and the "fever" [mostly malaria] incidence … which was in accordance with the air-borne theory of infection then generally accepted'. He was thus greatly influenced by the prevalence of malaria, which was clearly dependent on the magnitude of the recent monsoon, and hence the extent of stagnant water.

Regarding smallpox (variola), he found a prognostic relationship between 'saturation deficiency' and incidence of the disease, this being a 'much stronger predictive factor than *absolute* humidity'. In the Punjab, 'mean temperatures and saturation deficiencies … were very favourable to the subsequent increased plague prevalence. Regarding the 'most important epidemic disease in India' – cholera – Rogers again found it possible to predict epidemics from climatic factors.[31] Thus Rogers' meticulous work in eastern India is a reminder of a former era, prior to the 'germ theory' of disease (see Chapter 1), when climatic factors were deemed crucially important in the prediction of epidemics in the tropics.

NOTES

1 Anonymous. Sir Leonard Rogers. *Nature, Lond* 1938, 14 May; L Rogers. *Happy Toil: fifty-five years of tropical medicine*. London, 1950: Frederick Muller Ltd, p. 271; C Singer, E A Underwood. *A Short History of Medicine*, 2nd edn. Oxford, 1962: Clarendon Press, pp. 493–4; P Manson-Bahr. Major-General Sir Leonard Rogers. In: *History of the School of Tropical Medicine in London (1899–1949)*. London, 1956: H K Lewis, pp. 149–50; Anonymous. Leonard Rogers. *Lancet* 1962, ii: 666–7; Anonymous. Sir Leonard Rogers. *Br Med J* 1962, ii: 862–3, 932, 1200; J S K Boyd. Leonard Rogers 1868–1962. *Biographical Memoirs of Fellows of the Royal Society*, Vol. 9. London, 1963: The Royal Society, pp. 261–85; Anonymous. Rogers, Sir Leonard. *Munk's Roll*, Vol. 5. London: Royal College of Physicians, pp. 353–5; G McRobert, H J Power. Rogers, Sir Leonard (1868–1962). In: H C G Matthew, B Harrison (eds), *Oxford Dictionary of National Biography*, Vol. 47. Oxford, 2004: Oxford University Press, pp. 572–4; G C Cook. Leonard Rogers, KCSI, FRCP, FRS (1868–1962) and the founding of the Calcutta School of Tropical Medicine. *Notes Rec R Soc Lond* 2006, 60: 171–81.

2 G C Cook. William Twining (1790–1835): the first accurate descriptions of 'tropical sprue' and kala-azar? *J Med Biog* 2001, 9: 125–31.

3 L Rogers. *Fevers in the Tropics*. Oxford, 1908: Oxford Medical Publications.

4 L Rogers. *Report of an Investigation of the Epidemic of Malarial Fever in Assam, or Kala-azar*. Shillong, 1987: Assam Secretariat Printing Office, p. 223; G C Low. A retrospect of Tropical Medicine from 1894 to 1914. *Trans R Soc Trop Med Hyg* 1929, 23: 213–32.

5 *Op cit.* See note 1 above (Rogers). See also: *Op cit.* See note 4 above (Low); H H Scott. Amoebic dysentery and hepatitis. In: *A History of Tropical Medicine*. London, 1939: Arnold, pp. 820–39; R S Bray. Amoebiasis. In: F E G Cox (ed.), *The Wellcome Trust Illustrated History of Tropical Diseases*. London, 1996: Wellcome Trust, pp. 170–77.

6 G C Cook. Management of cholera: the vital role of rehydration. In: B S Draser, B D Forest (eds), *Cholera and the Ecology of* Vibrio cholerae. London, 1996: Chapman & Hall, pp. 54–94.

7   L Rogers. Presidential Address. The methods and results of forecasting the incidence of cholera, smallpox and plague in India. *Trans R Soc Trop Med Hyg* 1933–4, 27: 217–38; *Op cit.* See note 3 above (Rogers).

8   *Op cit.* See note 1 above.

9   Alfred McCabe-Dallas studied medicine at Guy's Hospital, graduating LSA Lond in 1900. He then served as an Assistant Medical Officer at the Boro' Asylum Portsmouth. He was a Civil Surgeon to the South Africa Field Force from 1901 to 1902, and then took the DTM diploma at The Liverpool School of Tropical Medicine before proceeding to Assam. See also Anonymous. *Medical Directory.* London, 1920: J & A Churchill, p. 839.

10 Sir (Samuel) Rickard Christophers FRS (1873–1978) was a leading protozoologist and specialist in Tropical Medicine. He was educated at University College, Liverpool, and proceeded to South America and India (where he served in the research branch of the Indian Medical Service). He was subsequently Professor of Malarial Studies at the London School of Hygiene and Tropical Medicine. See also C Garnham. Christophers, Sir (Samuel) Rickard (1873–1978). In: H C G Matthew, B Harrison (eds), *Oxford Dictionary of National Biography*, Vol. 11. Oxford, 2004: Oxford University Press, pp. 559–60.

11 John William Watson Stephens FRS (1865–1946) was an eminent parasitologist and tropical diseases expert. He received his training at Gonville and Caius College, Cambridge, and St Bartholomew's Hospital, London. After service in India and British Central Africa, Stephens spent most of his career at the Liverpool School of Tropical Medicine, where he became Walter Myers Professor of Tropical Medicine. See also W F Bynum. Stephens, John William Watson (1865–1946). In: H C G Matthew, B Harrison (eds), *Oxford Dictionary of National Biography*, Vol. 52. Oxford: Oxford University Press, pp. 476–7.

12 A McCabe-Dallas. Tropical School of Medicine for India. *Englishman, Calcutta* 1910, 11 March: 10.

13 L Rogers (editorial). A Calcutta School of Tropical Medicine. *Englishman, Calcutta* 1914, 21 March: 4.

14 L Rogers. Proposed school of tropical medicine in Calcutta. *Br Med J* 1910, i: 1010.

15 G C Cook, A J Webb. The Albert Dock Hospital, London: the original site (in 1899) of Tropical Medicine as a new discipline. *Acta Tropica* 2001, 79: 249–55.

16 P J Miller. *'Malaria, Liverpool': an illustrated history of the Liverpool School of Tropical Medicine 1898–1998.* Liverpool, 1998: Liverpool School of Tropical Medicine, p. 78.

17 G C Cook. 'Tropical' cases admitted to the Albert Dock Hospital in the early years of the London School of Tropical Medicine. *Trans R Soc Trop Med Hyg* 1999, 93: 675–7.

18 Sir Charles Pardey Lukis (1857–1918) KCSI, FRCS received his medical education at St Bartholomew's Hospital, London. He entered the Bengal Army as a surgeon on 31 March 1880. Lukis became Civil Surgeon in Simla in 1899, and Hon. Surgeon to the Viceroy in 1905. He was also appointed Professor of Materia Medica at the Calcutta Medical College and, in 1905, Professor of Medicine and Principal of the College and first Physician of the Hospital attached to the College. He was ultimately promoted to the rank of Lieutenant-General on 22 September 1916. See also: Anonymous. Lukis, Sir Charles Pardey. *Plarr's Lives* 1930, i: 741–2; Anonymous. *Who Was Who 1916–1928*, Vol. 2. London, 1947: A & C Black, p. 651.

19 Sir Edward Norman Baker KCSI (1857–1913) was Lieutenant-Governor of Bengal from 1908 until 1911, and an ordinary member of The Council of India from 1905. He had been a member of the Bengal Legislative Council from 1900 until 1902, and Financial Secretary to the Government of India from 1902 to 1905. See Anonymous. *Who Was Who 1897–1916.* London, 1935: A & C Black, p. 35.

20 Anonymous. *A Short Account of the Calcutta School of Tropical Medicine, Institute of Hygiene and the Carmichael Hospital for Tropical Diseases.* Calcutta, 1922: Alliance Press, p. 31.

21 Henry Arthur Crouch (1870–1955) CIE, FRIBA received his early education at Brisbane Grammar School. After several architectural successes in England, he became Consulting Architect to the Government of Bengal (1909–35) and Consulting Architect to the Government of India, during which he designed the School of Tropical Medicine and Hospital for Tropical Diseases,

Calcutta. He also designed numerous other buildings in India. See also: Anonymous. *Who Was Who 1951–1960*. London, 1961: A & C Black, pp. 262–3.

22 Baron Carmichael of Skirling (1859–1926) was born in Scotland, his father was tenth baronet. He was educated at St John's College, Cambridge. Carmichael was appointed in May 1911 to the Governorship of Madras, but the following April he was appointed first Governor of the newly reconstituted state of Bengal. From his early days he had a great interest in art. See also K Prior. Carmichael, Thomas David Gibson Carmichael, Baron Carmichael (1859–1926). In: H C G Matthew, B Harrison (eds), *Oxford Dictionary of National Biography*, Vol. 10. Oxford, 2004: Oxford University Press, pp. 179–90.

23 Major-General Sir William Rice Edwards (1862–1923) KCB, KCIE, FRCSE was educated at Magdalen College School Oxford, Clifton College and the (Royal) London Hospital. After junior appointments at his teaching hospital, he entered the Indian Medical Service (IMS) in 1886. Following service in South Africa (with Lord Roberts), he held a number of administrative posts in India before becoming Director-General of the IMS (1918–22), including member of the Bengal Legislative Assembly (1915–18, the Government of India (1918–20), and the Council of State for India (1920–22). See Anonymous. *Who Was Who 1916–28*. London, 1947: A & C Black, p. 322.

24 Sir William Duke (1863–1924) was born in Scotland, and after joining the Indian Civil Service was posted to Bengal. In 1909 he became Chief Secretary for Bengal, and was ultimately in charge of the Medical Department. See P G Robb. Duke, Sir (Frederick) William (1863–1924). In: H C G Matthew, B Harrison (eds), *Oxford Dictionary of National Biography*, Vol. 17. Oxford, 2004: Oxford University Press, pp. 164–5.

25 Anonymous. Tropical School of Medicine: dispute over staffing: sensational development: possible resignation of Colonel Rogers. *Englishman, Calcutta* 1914, 4 April: 4.

26 Anonymous. Tropical School of Medicine: a Government Communiqué. *Englishman, Calcutta* 1914, 7 April: 7; Anonymous (editorial). The staffing dispute in the School of Tropical Medicine. *Ibid*. 1914, 7 April: 4.

27 Sir David Yule, Bt (1858–1928) was Director of the Midland Bank, the Mercantile Bank of India, Vickers Ltd, and the Royal Exchange Assurance. He was described as a 'shy and reclusive man'. In 1875, he went with his uncle to Calcutta to manage the Bengal Cotton Mills. See also: Anonymous. *Times, Lond* 1928, 4 July: 21; Anonymous. *Who Was Who 1916–1928*. London, 1947: A & C Black, p. 1158; Russell I F. Yule family. In: H C G Matthew, B Harrison (eds), *Oxford Dictionary of National Biography*, Vol. 60. Oxford, 2004: Oxford University Press, pp. 973–4.

28 H Power. The Calcutta School of Tropical Medicine: institutionalizing Medical Research in the Periphery. *Med Hist* 1996, 40: 197–214.

29 Anonymous. *Annual Report of the Calcutta School of Tropical Medicine, Institute of Hygiene and the Carmichael Hospital for Tropical Diseases for the year 1922*. Calcutta, 1923; Alliance Press, p. 67. Anonymous. *Ibid. 1923*. Calcutta, 1924: Lal Chand & Sons, p. 86.

30 *Op cit*. See note 28 above.

31 L Rogers. The methods and results of forecasting the incidence of cholera, smallpox and plague in India. *Trans R Soc Trop Med Hyg* 1933, 27: 217–38.

14

Aldo Castellani (1877–1971): research in the tropics, and founding of the Ross Institute and Hospital for Tropical Diseases

Aldo Castellani (1877–1971; Figure 14.1) was born in Florence, Italy, on 8 September 1877. Education at Florence, where he graduated in medicine in 1899, and Bonn (under Profession W Kruse) followed, and after this Louis Sambon (1865–1931), a medical graduate of the University of Naples, suggested that he apply to work in Manson's Department at the London School of Tropical Medicine (LSTM), then situated at the Albert Dock Hospital. He duly arrived in London in 1901, and was soon appointed (with G C Low and C Christy) to the first Royal Society Sleeping Sickness Expedition of 1902–3 (see Chapter 8). Castellani claimed priority for a causal relationship with trypanosomes, and the matter subsequently led to a great deal of acrimonious correspondence (see Chapters 8 and 9).[1]

Following this relatively unsuccessful expedition, Castellani was appointed bacteriologist (at Colombo) to the Government of Ceylon (now Sri Lanka). He remained there for twelve years (1903–15), during which time he discovered the causative agent of yaws (*Treponema pertenue*), in 1906; he was also able to demonstrate transmission of that infection to a subhuman primate. Castellani was one of the first physicians to inject salvarsan (arsphenamine), recently elaborated by Ehrlich (1854–1915) as a cure for yaws. Entering the relatively unexplored field of mycology, in 1907 he isolated the organism responsible for *tinea imbricata*; he also

FIGURE 14.1 Aldo Castellani FRCP (1877–1971) (reproduced courtesy of The Wellcome Library, London).

introduced carbol fuchsin (Castellani's) paint as a cure for mycotic infection(s). In 1912, he wrote, with A J Chalmers, the first edition of his textbook, *Manual of Tropical Medicine*.

In 1915, at the age of 38 years and already an established tropical physician, Castellani returned to Italy as Professor of Medicine at Naples. He subsequently became involved with the Red Cross in Serbia during the First World War (1914–18), during which he produced a (relatively ineffective) antityphus vaccine. In 1917, he was called to Paris as a member of the Inter-allied Sanitary Commission. He also at this time identified a free-living amoeba, *Hartmannella castellanii*.

Returning to London, unable to find suitable employment in Italy, he became consultant to the Ministry of Pensions at Roehampton, and was elected FRCP (having obtained the MRCP in 1916) in 1922. Castellani's Harley Street practice grew to astronomical proportions. Despite this, he taught in London from June to December, in New Orleans from January to March, and in Rome from April to June. He also edited the *Journal of Tropical Medicine and Hygiene* for many years.

Before the Second World War (1939–45), Castellani was summoned to Italy on several occasions to treat Benito Mussolini ('*Il Duce*') (1883–1945). He was later appointed Surgeon General to the Italian forces during the Italo-Abyssinian (Ethiopian) conflict; at this time he was closely associated with Mussolini and was created Count of Chisimaio, and in 1946 he was made a Marchese by King Umberto II. At the outbreak of war he again headed for Rome, and shortly afterwards directed medical operations in Tripoli and Cyrenaica as Director of Medical Services. In 1945 he lost his Chair in Rome, due to alleged Fascist tendencies, and settled in Cascais, near Lisbon, as a member and medical advisor of the Italian Royal household – then exiled to Portugal. For some years he was Professor of Pathology and Tropical Medicine at the Lisbon Institute for Tropical Diseases, where his research concentrated on mycotic infections of the bronchi and lungs. He remained in Portugal for the remainder of his long life.

In 1907 Castellani married Josephine Ambler Stead, by whom he had one daughter, who married Lord Killearn.[2] He died in Lisbon on 3 October 1971, and a memorial service was held at St Paul's Cathedral on 4 November of that year.[3]

Castellani received many honours from France (Knight Officer of the Legion of Honour), the USA (Distinguished Service Cross) and Britain (being created an honorary KCMG in 1928; although this award was rescinded in 1940, it was reinstated shortly before his death). His autobiography first appeared in 1960.[4]

THE ROSS INSTITUTE AND HOSPITAL FOR TROPICAL DISEASES

The following account of early meetings to establish a Ross Institute, instigated by Dr Aldo Castellani in 1917, was provided by Sir Ronald Ross FRS (see Chapter 5) in May 1926:[5]

> On looking up my old diaries which I have kept fairly regularly since 1914, I find that Dr Aldo Castellani first came to me with the suggestion that he should move in the matter of raising a subscription in order to found an Institute bearing my name, where I could continue research, on Thursday the 20th Sep. 1917. After considering the matter I decided against the scheme as, in fact, I was in such poor health that I doubted whether I was likely ever to carry on any more investigations, except perhaps in mathematics.
>
> After the [First World] War, on the 15th February 1922, Castellani came again to me with the same suggestion. He considered that, even if I was still unable to continue research, the Institute might do good work without me. Finally I agreed; and then eight days later, on the 23rd February 1922, he and Dr William Simpson [1855–1931] came again to me at Harley House, Marylebone Road, where my flat was, in order to consider details. The result was that they immediately set about forming a Committee & asking the public for funds. Their appeal was published in *The Times* on Friday 22nd June 1923.

From this handwritten résumé, there can be no doubt that Castellani was the first person to suggest a Ross Institute. The renewed initiative was almost certainly precipitated by a letter (from Ross) to *The Times* in early January 1922;[6] the major thrust of this communication was that studies on malaria dated back some 2000 years, and that no living person was cognizant of what had and had not been discovered about the infection. Therefore an 'Imperial Bureau' targeted solely at malaria should be established. Ross considered that the two Schools of Tropical Medicine (at London and Liverpool), together with their respective bureaux, and in addition one designed to record entomological matters, were at that time 'engaged upon a very wide subject, of which malaria [was] only one part'. In a letter to Ross, Castellani asked whether he should 'enter into communication with the Rockefeller people *re* – the Central Tropical Research Institute in London'. It was, he felt, more likely that funds would be forthcoming for an institution than 'for a simple malarial bureau'.

At this time, Castellani suggested as a title for the institute, 'Central Institute for Research in Malarial and other Tropical Diseases'.[7] In a letter two days later, however, he had clearly reverted to his original idea, and suggested 'The Ross Institute', to bring the potential foundation into line with the Lister, Pasteur, Kitasato and Koch Institutes; there could be no better name, he continued, 'for a Central Research Institute for Malaria and other Tropical Diseases' than this![8] While approving an approach to the 'Rockefeller people', Ross considered that a more appropriate title might be 'Central Bureau and Research Institute in Malaria and other Tropical Diseases'; he did not, he claimed, like the idea of a 'Ross Institute', although that 'would certainly be a great honour'.[9] Having ascertained that 'Prof Simpson [was] in favour of [the] scheme', Castellani arranged a visit to Ross's flat (36 Harley House, Marylebone Road, NW1) on 15 February. After a further visit, accompanied by Simpson, on 23 February, Castellani 'immediately set about forming a Committee asking the public for funds'.[10]

Initial negotiations

Over 12 months were to pass after the initial overtures until things really became active. In a letter requesting support for the venture to Mr Harry Gordon Selfridge (1858–1947), Castellani outlined the background to his proposed scheme:[11]

> Perhaps you know that Sir Ronald Ross is not attached to the [LSTM] or to any other scientific Institute – he has no laboratory and no beds anywhere. For several years he has accordingly been unable to continue his researches on various problems in Tropical Medicine. Although he has borne with great fortitude this neglect, still we know it has been a cause of great unhappiness to him.

Castellani also wrote to many other distinguished men, including Sir Hugh Clifford (1866–1941), Governor of Nigeria, and Sir Harry Johnston (1858–1927), diplomat and author. Several dinners involving Castellani and Simpson were arranged at the Oriental Club, Hanover Square, W1, in order to 'butter up' potential signatories. There was obvious concern at this stage regarding the possibility of '*competition* [my italics] with the [LSTM]'; Sir Joseph West Ridgeway (1844–1930) would, for example, only sign the letter to *The Times* (see below) provided such a conflict of interests was eliminated. Castellani wrote to Ross from 33 Harley Street, however, explaining, that 'there will be *no teaching* [my italics] and therefore there is no question of competition'.

An idea of public feeling from abroad towards the venture can be gained from a letter, addressed to Castellani, from the Surgeon General of the United States:

> I feel that you have conferred an honour upon me in giving me an opportunity to express my appreciation of Sir Ronald Ross, whom I have had the privilege of knowing personally.

To my mind the work of Ronald Ross placed him permanently in that small group of great men to whom not only the British Empire but the world at large owes a debt of gratitude which can never be fully re-paid. The work of my countrymen in Cuba and in Panama [i.e. the control of malaria and yellow fever by W C Gorgas, 1854–1920; see Chapter 6] would not have been possible but for the application of the fundamental truths discovered by him.

Other signatories of the appeal included Sir Humphry Rolleston, President of the Royal College of Physicians; Sir Anthony Bowlby, President of the Royal College of Surgeons; Professor Filippo Rho, President of the International Health Commission of the Danube and Professor of Tropical Hygiene at the Royal University of Naples; Sir Dorabji Tata; Sir Woodman Burbridge Bt (Managing Director of Harrods); the Marquess of Lansdowne; Lord Cable; Sir Jesse Boot Bt; and Professor J Bordet (of Belgium).[12]

The appeal letter was eventually published (with 33 signatories, headed by the Rt Hon H H Asquith (1852–1928), a former Prime Minister) in *The Times* of 22 June 1923. Among other well-known names were: Sir Edward Marshall-Hall KC, Sir James Cantlie, Sir Frederick Mott, Sir Thomas Horder Bt, Professor W H Welch, Sir Arbuthnot Lane Bt, and Sir Charles Sherrington (President of the Royal Society).[13] Congratulatory messages to Ross included those from A O P Reynolds, Henry Morris, and Sir James Chrichton-Browne. The last pointed out (Ross was renowned for seeking pecuniary reward for his 1897–8 discoveries) that Sir Frederick Banting (1891–1941) had just been granted £14000 annually by the Canadian Parliament for his 'discovery of the value of insulin in diabetes'; diabetes, he considered, was a mere 'flea-bite compared with malaria, and insulin is not a preventive nor a cure'.[14]

Criticism of Castellani's initiative

Not all reaction to the appeal was congratulatory, however. Senior figures in both the London and Liverpool Schools of Tropical Medicine feared that the proposed institute would overlap significantly with their already well-established activities. The Dean of the LSTM (at that time situated at Endsleigh Gardens), together with his colleagues, objected to a statement that 'the range of activities [of the School] is in the main limited to training young medical men for the practice of their profession in the tropics'. This statement totally ignored, they claimed, its research component! The same theme was echoed by a representative of the Liverpool School; numerous expeditions to tropical countries, which must have been very well known to Ross, were outlined, as was the scientific work accomplished in the Liverpool laboratories.[15] These letters brought forth a rapid response from Sir West Ridgeway (one of the signatories of *The Times* appeal letter; there was, he concluded, 'no reason whatever why the three institutions should not work together in cordial harmony for the benefit of mankind'. This letter in turn brought forth a speedy reply from the Dean (and his colleagues) of the LSTM; he pointed out that 'ever since its inception … as one of the institutions of the Seamen's Hospital Society (SHS) [it had] consistently utilized the [research] opportunities thus afforded by the hospitals with which it has been associated'. Sir James Barr, who was *not* a signatory of the appeal letter, also wrote to *The Times*, commending the proposed 'permanent' memorial to Ross, who he considered deserved 'an earldom with £100 000 to enable him to maintain his well-merited dignity'. He considered that, apart from Sir Leonard Rogers' (see Chapter 13) work, no original discovery had emanated from the tropical schools; he could, he wrote, 'almost hear these men crying out "Ye men of Ephesus, Our craft is in danger!"'[16]

Formation of the Ross Institute

On 29 June 1923, the first meeting of the Executive Committee of what was then known as 'The Ross Clinique for Tropical Diseases' was held at the Institute's temporary offices at 56 Queen Anne Street, W1. In addition to Castellani and Simpson (the Organizing Committee), Sir James Cantlie, Sir Allan Perry, W Shakspeare and W G Partington were present. Simpson was elected Chairman and Partington Honorary Secretary; the fact that Castellani was the originator of the scheme was highlighted. Numerous other individuals were elected to serve on the permanent Executive Committee, which was to meet 'about every fortnight', of which Ross was to be Chairman; Ross, Perry and Colonel W G King (1851–1935) were elected to the (expanded) Organizing Committee, which from then onwards was to be called 'The Scientific Organizing Committee'. 'Finance' and 'Press and Propaganda' Committees were also established and members elected.

The fact that the Prince of Wales had subscribed £10 to the Institute (the appeal was launched to raise £50 000) was recorded in *The Times* for 13 August 1923. In the opinion of the Poet Laureate, John Masefield (1878–1967) – a close friend of Ross – writing in a foreword to a 'short pamphlet' issued by the 'Ross Research Institute', Ross's discovery had 'cut the Panama Canal and made a third of the world habitable'. He considered:[17]

> This is the greatest thing done in our time by one man. The nation which produced that man should crown his deed with a living power to make his work not a memory, but a lifting up of life throughout the world.

A notice in *The Times* of 4 October the same year by the 'Paris Correspondent' drew attention to the death of M. Charles de Lesseps; the writer outlined the reasons for the 'tragedy of Panama' (see Chapter 6), but failed to mention the parts played by malaria and yellow fever. This serious omission was corrected two days later in a joint letter from Castellani and Simpson.[18]

In early December 1923, King, who had extensive experience of India and the Indian Medical Service (IMS), wrote to Ross enclosing various valuable (albeit somewhat complicated and verbose) suggestions regarding fund-raising strategies at home and in India. They were really intended for the Organizing Secretary, Major H Lockwood Stevens, to whom full details were supplied. One suggestion – that Masonic lodges should be involved in the appeal – was, he said, 'in my opinion ... inappropriate'. Another remark is worth reporting: 'The assumption of the connection between malaria and goitre being held may be correct in regard to laymen, but it is certainly a theory not held by medical men in Burma'. And with reference to Ross's resignation from the IMS, King considered: 'His freedom from official control ... necessitated his acceptance of a pension of about one-fourth the amount he would have earned had he continued to serve in India'. He was in favour of a 'Ross Applied Hygiene Federation [to attain ... international interchange of information which should prove of great advantage to humanity] being formed [which would include] all original subscribers to the Institute'.[19]

Meanwhile, Malcolm Watson (see Chapter 17)[20] had written to Ross, enclosing several articles he had written in the *Straits Times*, the *Penang Gazette*, the *Malay Mail* and the *Times of Malaya and Planters' & Miners' Gazette* for October/November 1923, from Klang (Kelang, Malaysia) giving details of energetic efforts he was making with regard to local fund-raising on behalf of the Ross Institute. He suggested in his letter getting in touch with the Rubber Growers' Association, and the colonial governments. In reply, Ross said how grateful he was for all that Watson was doing for the Institute, and to Lady Durning-Lawrence, who had donated sufficient cash (£1000) for him (Ross) to stay in London for a further two years: 'I do not know what will happen after that either if I am alive or dead! Perhaps in the latter event St Peter will be more charitable than some people have been'. In a letter thanking a Mrs E E Holt for her subscription, Ross alluded again to his poor state of health during 1923: 'both my wife and self have been ill off and on for nearly the whole year and she had acute bronchitic pneumonia and has been dangerously ill'.[21]

A suitable 'home' for the Institute

By early 1924, an urgent appeal (from the temporary headquarters at 56 Queen Anne Street) for funds to purchase a building for the Ross Institute was underway; a letter appealing for funds – the cost of a suitable property would be £30000 – was duly published in *The Times* for 26 January, and this was signed by the Duchess of Portland (Vice-President, RIHTD), Lord Hardinge (a former Viceroy of India), Sir John Hewett, Sir Arbuthnot Lane and William Shakspeare. This communication had, however, been preceded by considerable lobbying; for example, Sir Charles Sherrington was approached but, in view of the urgency of the matter and he being conscious of the fact that he should consult the Royal Society Council before signing the letter, reluctantly declined.[22]

But where was the suitable property to be? Simpson wrote to Ross a few days before *The Times* letter appeared, emphasizing the urgency involved in this difficult matter. He had, he claimed, 'visited scores of freehold and long-lease houses without any grounds, and there [was] none suitable except at exorbitant prices'.[23]

The first realistic possible venue to appear on the market was Northgate House, Avenue Road, Regent's Park, which was advertised freehold – a building surrounded by two acres of ground, for £20 000. Ross immediately wrote to Lord Leverhulme (1851–1925), President RIHTD, to tell him about this property and to ask for his financial support. In a letter to Ross, Leverhulme offered a 'cash payment of £1000' together with an assurance of guarantor with others for a further £1000 – 'the amount of my guarantee not to exceed £1000 and to be a personal guarantee and not jointly and severally'. The offer of £20 000 for the freehold of Northgate House was accepted; according to Ross, they had £10 000 available and the remainder would be 'found by mortgage'. In his letter to Leverhulme, Ross had asked him to be one their trustees; the others were to be himself and Simpson. As the originator of the project, it seems slightly odd that Castellani was not nominated to fill one of these positions. In a letter to Lane telling him of the purchase, Ross asked whether he was going on a Central American expedition; Ross had decided not to go, owing to his 'diabetic tendency'. In a further letter to Leverhulme, Ross reported that, owing to various clauses *in the lease* (my italics) preventing use as a hospital, negotiations for the Regent's Park property had fallen through.[24]

Two other houses then came into the reckoning: Bellmore on Hampstead Heath, and 2 Chelsea Embankment. In a subsequent letter to an estate agent, Ross showed an interest in Cheyne House, 18 Chelsea Embankment (which would be most suitable, 'because we can doubtless arrange to bring our few patients by launch from the docks') – and outlined the objectives for the Ross Institute: a small hospital for patients not suffering from diseases which are 'communicable in this country', together with a research institute in which will be kept 'some small animals, such as rabbits and guinea pigs'. The Institute, he continued, would have the same object as the Lister Institute on the Thames Embankment, the Pasteur Institute in Paris, and many other research institutions. He gave his assurance that the proposed Institute 'will be no nuisance' to neighbouring houses. Unfortunately a 'restriction covenant' precluded use of this property, and this possibility ultimately fell through. In a letter to Ross dated later that month (September), Leverhulme said he was sorry that this property had been lost to the Institute, but in his opinion it was in any case no more central than Bellmore, Hampstead Heath, which he considered to be worth £12 000.[25]

In a personal letter to a Mr William Mackinnon of Aberdeen, Ross mentioned that they had just 'offered £15 000 for a fine property near Wimbledon' (this was Bath House, Putney Heath, which ultimately became the headquarters of the RIHTD); it was intended to have about 20 hospital beds and in addition the staff would 'investigate some home maladies such as cancer, which are closely related to many tropical diseases!'[26]

Meanwhile, the ambitious Castellani, originator of the project, was becoming increasingly unhappy with these delays. In a letter to Ross, he wrote: 'While I consider it a very great honour to be one of the Directors of the Institute, I may perhaps be allowed to mention that other appointments have been offered to me'. He had in fact been offered a Professorship (and Directorship) of the new Tropical Medicine School at New Orleans.[27]

Role of the Seamen's Hospital Society

In early 1925, the Secretary of the Seamen's Hospital Society, Sir James Michelli (1853–1935), wrote to the RIHTD indicating that his Society was unhappy with the title of the Institute; their present establishment at Endsleigh Gardens, WC1, was already called the Hospital for Tropical Diseases (HTD)! This correspondence was discussed at a meeting of the Organizing and Finance Committee of the RIHTD held on 13 January 1925, apparently without any clear decision being taken. Incidentally, at that meeting the surveyor's report and draft agreement regarding Bath House (see Figure 14.2) were also discussed and presumably approved. The project went ahead, however, with Ross as Director-in-Chief, Castellani as Director of Tropical Medicine and Dermatology, and Simpson (Figure 14.3) as Director of Tropical Hygiene.

The precise role of Castellani remains unclear; it seems that the 'hospital' component of the Institute was used by him as a small 'nursing home' to extend his already substantial private practice. The 'researches' of the Institute (mainly by Castellani) were to produce very mediocre results. At a meeting of the Executive Committee held on 27 January 1926, it was decided to appoint Stevens to the Secretaryship of the Institute and to dispense with the services of Frederick Hornyik (until then Secretary RIHTD), who was given 'three months salary in lieu of notice'.[28] The reason for his abrupt dismissal shortly before the official opening is not stated; however, a recently published article might provide a clue![29]

The completed institution

The RIHTD was officially opened by HRH the Prince of Wales on 15 July 1926.

The resultant product was never a great success. Neither clinical nor research components could compete with the already well-established facilities at the HTD and the newly-opened London School of Hygiene and Tropical Medicine (LSHTM).[30]

CONCLUSIONS

In retrospect, Castellani's most important contribution to British tropical medicine lay not in his research but in the textbook which he wrote (jointly), and which went to three editions. There seems little doubt that he was acting in good faith in instigating and founding (with Sir William Simpson) the RIHTD, although it might be suspected that his *raison d'être* was to some extent sycophantic – Ross was, after all, a major figure in early-twentieth-century British medicine.

FIGURE 14.2   Interior views of The Ross Institute and Hospital for Tropical Diseases, situated at Putney West Hill between 1926 and 1934 (reproduced courtesy of The Ross Institute Archive). For a view of the exterior, see Chapter 5.

It was widely appreciated in the 1920s that London did not require a second tropical institution, from the viewpoint of either patient care (the clinical interests were being well looked after by physicians at the SHS-run HTD or (what is more relevant in this context) research. Ross, as well as Castellani, must have

FIGURE 14.3   Afternoon tea-party at the RIHTD: Sir William (extreme right) and Lady Simpson (second from left) are shown, together with Major Lockwood Stephens and the matron (reproduced courtesy of The Ross Institute archive).

been aware of this! What was required was a preventive medicine institution to give advice to individuals working in the tropics, but that would undoubtedly have overlapped with the activities of the newly-established LSHTM, which had been founded in 1924 and was formally opened in 1929 (i.e. during the lifetime of the RIHTD). There would, of course (as Ross must also have been only too well aware), be overlap between the envisaged activities of the RIHTD and the Liverpool School of Tropical Medicine. The RIHTD in fact collapsed (its financial standing was never satisfactory) after Ross's death on 16 September 1932.[31]

The research institute became (after lengthy negotiations in which Castellani proved less than helpful) incorporated into the LSHTM, and four beds were established in a 'Ross Ward' at the HTD in 1934.[32]

NOTES

1   G C Cook. Correspondence from Dr George Carmichael Low to Dr Patrick Manson during the first Ugandan Sleeping Sickness Expedition. *J Med Biog* 1993, i: 215–29; Castellani, Professor Marchese Count Aldo. In: *Who Was Who 1971–80*, Vol. 7. London, 1981: A & C Black, p. 135. G C Cook. Aldo Castellani FRCP (1877–1971) and the founding of the Ross Institute & Hospital for Tropical Diseases at Putney. *J Med Biog* 2000, 8: 198–205.

2   P Manson-Bahr. In: *History of the School of Tropical Medicine in London (1899–1949)*. London, 1956: H K Lewis, pp. 255–7; A Castellani. *Microbes, Men and Monarchs: A Doctor's Life in Many Lands*. London, 1963: Gollancz, p. 287; Anonymous. Aldo Castellani. *Lancet* 1971, ii: 883;

Anonymous. Aldo Castellani. *Br Med J* 1971, ii: 175, 307; Anonymous. Professor Marchese Sir Aldo Castellani: a distinguished physician. *Times, Lond* 1971, 5 October: 16; P C C Garnham, G Wolstenholme. Aldo (Count of Chisimaio) Castellani. In: G Wolstenholme (ed.), *Munk's Roll*, Vol. 5. London: Royal College of Physicians, pp. 92–4.

3 Anonymous. Memorial service. Professor Marchese A Castellani. *Times, Lond* 1971, 5 November: 17.

4 *Op cit.* See note 1 above (Castellani).

5 R Ross. *Memoirs, With a Full Account of the Great Malaria Problem and Its Solution.* London, 1923: John Murray, p. 547; E R Nye, M E Gibson. *Ronald Ross: Malariologist and Polymath: A Biography.* London, 1997: Macmillan, p. 316; R Ross. *Original Suggestion of the Ross Institute.* Manuscript, 11 May 1926. Ross Archive, London School of Hygiene and Tropical Medicine. Dr W J R (later Sir William) Simpson was a physician and pioneer of tropical medicine. A graduate of Aberdeen University, he later became Medical Officer of Health for Calcutta and Professor of Hygiene at King's College, London. See also G C Low. William John Ritchie Simpson. *Br Med J* 1931, ii: 3 October.

6 R Ross. War on malaria. Need for Imperial Bureau. Sir Ronald Ross's scheme. *Times, Lond* 1922, 2 January: 11.

7 A Castellani to R Ross. 3 February 1922. Ross Archive, London School of Hygiene and Tropical Medicine.

8 A Castellani to R Ross. 5 February 1922. Ross Archive, London School of Hygiene and Tropical Medicine.

9 R Ross to A Castellani. 6 February 1922. Ross Archive, London School of Hygiene and Tropical Medicine; R Ross to A Castellani. 9 February 1922. Ross Archive, London School of Hygiene and Tropical Medicine.

10 A Castellani to R Ross. 11 February 1922. Ross Archive, London School of Hygiene and Tropical Medicine.

11 A Castellani to H G Selfridge. 10 March 1923. Ross Archive, London School of Hygiene and Tropical Medicine.

12 *Op cit.* See note 1 above (Cook).

13 H H Asquith *et al.* Tropical diseases. The debt to Sir Ronald Ross. Proposed institute as monument. *Times, Lond* 1923, 22 June: 15; Leading article. A Ross Institute. *Times, Lond* 1923, 22 June: 15.

14 *Op cit.* See note 1 above (Cook).

15 R H Charles *et al.* Tropical medicine. London and Liverpool Schools. *Times, Lond* 1923, 30 June: 8; J W W Stephens. Tropical medicine. London and Liverpool Schools. *Times, Lond* 1923, 30 June: 8.

16 W Ridgeway. Proposed Ross Institute. *Times, Lond* 1923, 7 July: 11; R H Charles *et al.* Tropical medicine: research in London. *Times, Lond* 1923, 20 July: 8; J Barr. A Ronald Ross Institute. *Times, Lond* 1923, 21 July: 15.

17 Minutes of meeting held by the Executive Committee of the Ross Clinique for Tropical Diseases. 29 June 1923. Ross Archive, London School of Hygiene and Tropical Medicine; Anonymous. Tropical disease inquiry. *Times, Lond* 1923, 13 August: 5; Anonymous. 'War against malaria': Ross research Institute's pamphlet. *Times, Lond* 1923, 21 September: 7.

18 Anonymous. Death of M. Charles de Lesseps. The tragedy of Panama. *Times, Lond* 1923, 4 October: 11; W J Simpson, A Castellani. The tragedy of Panama. *Times, Lond* 1923, 6 October: 6.

19 W G King to R Ross. 9 December 1923. Ross Archive, London School of Hygiene and Tropical Medicine; W G King. Memo 9 December 1923: 9. Ross Archive, London School of Hygiene and Tropical Medicine; W G King. Memo addressed to the Organizing Committee 31 December 1923: 3. Ross Archive, London School of Hygiene and Tropical Medicine; Anonymous. Death of Colonel King: former Madras surgeon. *'Hindu', Madras* 1925, 5 April.

20 (Sir) Malcolm Watson (1873–1955) was a malariologist with nearly 30 years' experience in Malaya. He was later to become Director of Tropical Hygiene and Principal of the Malaria Department of the RIHTD; subsequently, in 1934, he moved to the newly established Ross Institute at the London School of Hygiene and Tropical Medicine.

21 M Watson to R Ross. 12 November 1923. Ross Archive, London School of Hygiene and Tropical Medicine; R Ross to M Watson. 20 December 1923. Ross Archive, London School of Hygiene and Tropical Medicine; R Ross to E E Holt. 21 December 1923. Ross Archive, London School of Hygiene and Tropical Medicine.

22 W Portland *et al*. Tropical Diseases. Proposed premises for Ross Institute. *Times, Lond* 1924, 26 January: 6; C S Sherrington to R Ross. 23 January 1924. Ross Archive, London School of Hygiene and Tropical Medicine.

23 W J Simpson to R Ross. 20 January 1924. Ross Archive, London School of Hygiene and Tropical Medicine.

24 Leverhulme to R Ross. 4 May 1924. Ross Archive, London School of Hygiene and Tropical Medicine; R Ross to Leverhulme 27 May 1924. Ross Archive, London School of Hygiene and Tropical Medicine; R Ross to A Lane. 24 June 1924. Ross Archive, London School of Hygiene and Tropical Medicine; R Ross to Leverhulme. 16 June 1924. Ross Archive, London School of Hygiene and Tropical Medicine.

25 R Ross' Secretary to Miss Granville (Ross Institute), 17 July 1924. Ross Archive, London School of Hygiene and Tropical Medicine; R Ross to Hampton and Sons, Estate Agents and Auctioneers. 1 September 1924. Ross Archive, London School of Hygiene and Tropical Medicine; R Ross to J T Jefferies, Chelsea Town Clerk, 17 September 1924. Ross Archive, London School of Hygiene and Tropical Medicine; H Chick, M Hume, M MacFarlane. *War on Disease: A History of The Lister Institute*. London, 1971: André Deutsch, p. 251; Leverhulme to R Ross. 22 September 1924. Ross Archive, London School of Hygiene and Tropical Medicine.

26 R Ross to W Mackinnon. 22 October 1924. Ross Archive, London School of Hygiene and Tropical Medicine.

27 A Castellani to R Ross. 10 November 1924. Ross Archive, London School of Hygiene and Tropical Medicine; Anonymous. Study of tropical medicine. New Orleans post offered to Dr Castellani. *Times, Lond* 1924, 19 November: 13.

28 R Ross to P J Michelli. 6 January 1925. Ross Archive, London School of Hygiene and Tropical Medicine; Minutes of meeting of the Executive Committee of the RIHTD held on 27 January 1926. Ross Archive. London School of Hygiene and Tropical Medicine.

29 L Wilkinson, D J Bradley. A note on the early history of the Ross Institute. *Med Hist* 2001, 45: 507–10.

30 G C Cook. A difficult metamorphosis: the incorporation of the Ross Institute & Hospital for Tropical Diseases into the London School of Hygiene and Tropical Medicine. *Med Hist* 2001, 45: 483–506.

31 G C Cook. *From the Greenwich Hulks to Old St Pancras: a history of tropical disease in London*. London, 1992: Athlone, p. 338.

32 *Op cit*. See note 30 above.

15

Neil Hamilton Fairley (1891–1966): medicine in the tropics, and the future of clinical tropical medicine

Neil Hamilton Fairley (1891–1966; Figure 15.1) was born on 15 July 1891 in Melbourne, Australia, and educated in Melbourne at Scotch College and Melbourne University. At the outbreak of the Great War (1914–18) Fairley enlisted with the Australian Army Medical Corps, and from 1916 to 1918 he served in Egypt (where he contracted schistosomiasis) and in Palestine, achieving the rank of Lieutenant-Colonel. He worked on the immunology of schistosomiasis and malaria-associated anaemia. From 1920 to 1922 Fairley was first assistant to the Director of the Walter and Eliza Hall Institute of Research at Melbourne, where he worked on the serological diagnosis of hydatid disease and liver fluke infection.

From 1822 to 1825 he was Professor of Tropical Medicine at the University of Bombay; there he worked on (and suffered from) tropical sprue; he replaced the milk diet with a high protein one.

In 1929, Fairley was appointed to the consultant staff of the Hospital for Tropical Diseases (HTD), London. He undertook several visits to Macedonia, continuing his research on malaria-associated anaemia.

When the Second World War started, in 1939, Fairley enlisted with the Australian Medical Force, and soon began work on the prophylaxis and treatment of malaria. Using human volunteers at Cairns, he was able to demonstrate

FIGURE 15.1   Sir Neil Hamilton Fairley (1891–1966) (reproduced courtesy of The Wellcome Library, London).

the value of mepacrine and, later, proguanil in prophylaxis. This allowed allied troops to operate successfully in the Pacific, India and Burma (now Myanmar).

Following the war, Fairley returned to the HTD, and in 1946 he was appointed to the Wellcome Chair of Clinical Tropical Medicine at the London School of Hygiene and Tropical Medicine. Two years later, however, he suffered a stroke. Although he retired from the Chair, he continued most other work.

Fairley was elected FRS in 1942 and knighted in 1950. He also had numerous other honours bestowed upon him.[1]

FAIRLEY'S TROPICAL RESEARCH

Schistosomiasis

Schistosomiasis was an important medical condition in Egypt that affected soldiers in the First World War (1914–18); the full life-cycle of *Schistosoma haematobium* and *S. mansoni* had only recently been delineated (see Chapter 10).

Acute schistosomiasis (Katayama syndrome) caused a great deal of diagnostic confusion with other acute clinical entities, and, in order to facilitate early recognition, Fairley developed a complement-fixation test (along the lines of the Wasserman reaction for syphilis). The antigen was a saline and, later, an alcoholic extract of dried and powdered livers of infected snails. This test proved of enormous value in differentiating this acute disease from other conditions, such as typhoid and typhus, and also in monitoring treatment.

Fairley also endorsed Christopherson's demonstration of the efficacy of tarter emetic, but showed that only worms actually in the vascular system were killed by this agent. He also demonstrated the pathological effects of both species in sub-human primate infections.

While he was in India (1922–5), there was concern that troops returning from an infected area (such as Egypt) might introduce infection. He therefore explored the possibility of using cerceriae of the bovine schistosome as an antigen for a diagnostic test; a focus of infection with this disease (in water buffalo and other domestic animals) was known to exist near Bombay. He demonstrated that although emetine was effective at high dosage, this was unsuitable for human use, intravenous tartar emetic being the drug of choice.[2]

For further contributions to schistosomiasis research, see Fairley's Presidential Address to the Royal Society of Tropical Medicine & Hygiene (RSTMH) (see below).

Tropical sprue

During three years in Bombay, Fairley devoted a great deal of time to researching tropical sprue – which he personally contracted. As a result, he had to be invalided to a temperate climate; in fact he returned to the Walter and Eliza Hall Research Institute in Australia. Most observations were of a clinical nature. He showed that occasional tetany was caused by calcium malabsorption (and *not* parathyroid deficiency), and further that accompanying megaloblastic anaemia had nothing to do with Addisonian pernicious anaemia. Fairley also showed that the 'flat' glucose curve resulted from glucose malabsorption, and was not due to any form of endocrine deficiency.

He concluded that tropical sprue was fundamentally an intestinal disease, which was more common in tropical than temperate environments; most symptoms resulted from malabsorption of 'nutritive substances and vitamins'.

Fairley advocated a high protein diet, which was low in carbohydrate and fat; he in fact designed a commercially available ('Sprulac') diet based on these criteria.[3]

Malaria research

The anaemia consequent upon 'blackwater fever' was a subject of considerable interest to Fairley, as was nutritional macrocytic anaemia.

In 1941, as a Colonel in the Australian Army, Fairley investigated malaria in Macedonia, and, after claiming that non-immune British and Australian troops should not be sent there, advised that their deployment should instead be to Greece – where the danger(s) of malaria were much reduced.

In 1942 Fairley served briefly in Java and, following his return to Australia (and incidentally promoted to Brigadier) realized that the chichona plantations of Java had fallen into Japanese hands, while current supplies of quinine were rapidly becoming depleted. He received, from British and USA sources, supplies of mepacine (yet to be proven as an effective chemoprophylactic). He then infected human volunteers (at Cairns, Queensland) with *Plasmodium falciparum* and *P. vivax*, and found several unknown facts about the human malaria life-cycle; furthermore, mepacrine was shown to be an effective chemoprophylactic against *P. falciparum* infection. However, with *P. vivax* infection a relapse occurred soon after mepacrine was withdrawn. These observations had, of course, enormous implications for the 'war effort', and Boyd has concluded that this was 'Fairley's greatest contribution to medical science and his greatest triumph'. Later work at Cairns also established the efficacy of proguanil ('Paludrine') and chloroqine.[4]

Later work, in London, established the exo-erythocytic cycle of *P. vivax* infection. Although Fairley was acting essentially in an advisory capacity, largely based on his Cairns researches, this work was in the main carried out by two other distinguished malariologists: P C C Garnham (1901–94) and H E Shortt (1887–1957).[5]

Other research into diseases common in warm climates

Fairley, at various times, also carried out research on typhus fever, dracontiasis and other helminthic infections, snake-bite, filariasis, leptospiral jaundice (see Chapter 18) and bacillary dysentery – in which he demonstrated the value of sulphaguanidine in *Shigella* dysentery.[6]

FAIRLEY'S PRESIDENTIAL ADDRESS TO THE RSTMH

Fairley devoted his Presidential Address to the RSTHM in 1951 to *schistosomiasis* (see also Chapter 10) and problems associated with it. His reason for doing so, rather than concentrating on his more widely known malaria research, was that he had 'been associated with this field of research for 35 years, and also because of its increasing world-wide importance ...'. He began by highlighting the history of the infection, the Japanese contributions (see above), and Leiper's important researches in Egypt (see Chapter 10).

Fairley also referred to his researches using experimentally-infected monkeys. Diagnosis using a complement-fixation reaction, prevention, and treatment (as they stood in 1951) were also given prominence. He also referred to cattle schistosomiasis (*S. bovis* and *S. matthei* infection), and human susceptibility to these species.

He gave an excellent account of *S. japonicum* infection in south-east Asia involving USA and Australian troops during the Second World War. An analysis of ectopic lesions associated with schistosomal infection followed – this was not confined to *S. japonicum*, but also included *S. haematobium* and *S. mansoni* infections – and of complications including pulmonary involvement (Ayerza's disease), bladder carcinoma, and 'bilharzial splenomegaly with cirrhosis'. Diagnostic tests, including immunological ones (of which Fairley was a pioneer), and acquired immunity in man and monkeys also received in-depth coverage.

Fairley concluded his *tour de force* with a summary of the historical and contemporary states of chemotherapy in this important group of infections; this was of course several decades prior to the introduction of praziquantel, which is effective in all species of human schistosomiasis.[7]

INTEREST IN THE FUTURE OF *CLINICAL* TROPICAL MEDICINE

It should be noted that the terms of the Rockefeller bequest – which gave rise to the London School of Hygiene and Tropical Medicine (LSHTM) – did not provide for an associated hospital, and when this School was established, Manson's School and Hospital parted company.

Fairley has outlined discussions which took place during the war years (1939–45) regarding this subject. The HTD (at Endsleigh Gardens) was vacated at the outbreak of war; two out-patient sessions weekly were, however, continued there until 10 May 1941, when, after the explosion of a land-mine, the building was rendered unsafe. W E Cooke (see Chapter 19) continued the clinics at the LSHTM, while in-patient care was temporarily removed to the Dreadnought Hospital, Greenwich.[8]

But what should be done *after* the war? The feasibility of an Imperial Hospital in close proximity to the LSHTM had been seriously considered in the 1920s; this had been to some extent precipitated by the possibility that the Seamen's Hospital Society might 'pull out' of tropical medicine and leave the present HTD high and dry. In early 1941, informal discussions, involving the LSHTM and the Colonial Office, began regarding the possibility of establishing a new tropical hospital at Greenwich. By November of that year, the idea had been agreed in principle. However, Manson-Bahr (see Chapter 18), feeling responsible for the tenure of the HTD staff who were away during the war, was unable to associate himself with it. In May 1942, a continued undertaking by the SHS to the HTD staff added to the viability of this plan, and designs were drawn up by architects to build a Dreadnought Seamen's and Imperial Tropical Diseases Hospital at Greenwich. The Chairman (Harold MacMillan) of a meeting held at the Colonial Office backed the proposal wholeheartedly, and the way ahead seemed clear.

A serious hitch, however, occurred in April 1944, when the Ministry of Works refused to allow the Endsleigh Gardens building to re-open, and temporary clinical accommodation was established (a decision which was driven by the Colonial Office) at 23, Devonshire Street. This in consequence deflected impetus from the Greenwich plan.

A 'tropical hospital' in Bloomsbury

In March 1945, Fairley, back in London at the end of hostilities, wrote a memorandum suggesting that the Imperial Hospital should be established not at Greenwich but in Francis Fraser's (1885–1964) 'University City' in Bloomsbury[9] – in close proximity to the LSHTM. In January 1945, Fairley (at this time Hon Secretary of the RSHTM) submitted a detailed report to Council:

> Today no Hospital for Tropical Diseases exists in London, and the wing of the bomb-damaged [LSHTM], which was specially concerned with the teaching of parasitology, is no longer serviceable. ... a co-ordinated effort on the part of all interested bodies will be necessary to centralize teaching by building a modern Imperial Hospital for Tropical Diseases not far removed from the renovated [LSHTM] and the Wellcome Research Institution with its unique museum facilities ...

This report also contained an excellent overview of the state of tropical medicine throughout the world. Council regarded the matter of such great importance that it decided to bring it to the notice of the Secretary of State for the Colonies.

In a follow-up on 17 January 1946, 'Dr Macdonald said that the Colonial Office and some others were taking steps to provide beds at University College Hospital, but he knew nothing of the details of this scheme'. And at the next meeting of Council, on 21 February 1946, the minutes record that:

> In view of the various rumours abroad regarding plans for the establishment of a Tropical Hospital in London it did not now seem so urgent for the Tropical Medicine Policy Committee to meet. It might be better to wait till definite knowledge of what was being arranged became available ....

That this should be primarily a government and not an SHS enterprise was accepted in principle.

At this point, with the National Health Service (NHS) rapidly emerging, close links between the HTD at Devonshire Street and UCH were envisaged. It is not too surprising that association with an *undergraduate* teaching Hospital (tropical medicine would be controlled by the UCH Governing Body) completely failed to meet with approval from the HTD staff. Plans for the permanent centre in the 'University City' were therefore resurrected, but this would take time!

Fairley was meanwhile clearly of the opinion that Britain was now in serious danger of losing its position as the key player in *clinical* tropical medicine. The Chief Medical Officer, Wilson Jameson (1885–1962), stressed at this point that the Ministry of Health, and *not* the Colonial Office, would inevitably become responsible for funding the proposed institution under the NHS Act. The matter was subsequently brought to an end when the Rt Hon Aneurin Bevan (1897–1960) announced that, under the Act, the HTD would be linked with both UCH and St Pancras Hospital.

Fraser apparently remained firmly of the opinion, however, that at some future date the proposed Tropical Diseases Hospital and Centre at Bloomsbury (as proposed by Fairley) would subsequently come to fruition.[10]

WILL *CLINICAL* TROPICAL MEDICINE SURVIVE?

In my Presidential Address to the RSTMH in 1993, I warned (nearly 50 years after Fairley's involvement) that *clinical* tropical medicine was again in a state of turmoil; the 'Tomlinson report', published in October 1992, had recommended that the beds (and other services) at the HTD should move to UCH at last. Subsequent developments (with Mrs Virginia Bottomley, Secretary of State for Health, at the helm) rendered the future of that hospital, and indeed the NHS itself, extremely insecure. This situation has deteriorated even further under the present Government. I had written prophetically in 1992:

> at a meeting of the HTD Medical Committee held on 21 January 1992, the [decision to move the facilities to UCH Middlesex instead of St Thomas' – where an entire pavilion was assured] received the higher number of votes! In the judgement of this writer, if this goes ahead the identity of the HTD (the 'flagship' of the specialty, 'tropical medicine') is likely to be lost forever, and Manson's achievement (like that of the British Empire) will become a mere memory from a bygone era.

The future of tropical medicine as a discipline must inevitably be centred on the *clinical* component. Most tropical cases can indeed be dealt with by an infectious diseases physician, but a core of 'exotic' infections exists in which such a physician has no expertise whatsoever! It is essential, therefore, that the RSTMH supports wholeheartedly the continuation of this specialty – not only in London, but also elsewhere. The Liverpool School of Tropical Medicine currently seems a great deal more sympathetic to the *clinical* discipline than does the predominantly public health-dominated School in London.[11]

NOTES

1 P Manson-Bahr. Professor Sir Neil Hamilton Fairley. In: *History of the School of Tropical Medicine in London*. London, 1956: H K Lewis, pp. 152–5; Anonymous. *Times, Lond* 1966, 21 April; A W Woodruff. Sir Neil Hamilton Fairley. *Nature, Lond* 1966, 210: 1205; A W Woodruff. Neil Hamilton Fairley. *Lancet* 1966, i: 987–8, 1045; Anonymous. Sir Neil Hamilton Fairley. *Br Med J* 1966, i: 1117–18, 1178, 1366–7; J Boyd. Neil Hamilton Fairley (1891–1966). *Biographical Memoirs of Fellows of the Royal Society*, Vol. 12. London, 1966: Royal Society, pp. 122–41; Anonymous. *J Trop Med Hyg* 1973, 76: 233; A W Woodruff. Fairley, Sir Neil Hamilton. *Munk's Roll*, Vol. 6. London, Royal College of Physicians, pp. 171–3; J Boyd. Fairley, Sir Neil Hamilton

(1891–1966). In: H C G Matthew, B Harrison (eds), *Oxford Dictionary of National Biography*, Vol. 18. Oxford, 2004: Oxford University Press, pp. 948–9.

2 *Op cit.* See note 1 above (Boyd, 1966).

3 *Ibid.* See also G C Cook. Tropical Sprue. In: F E G Cox (ed.), *The Wellcome Trust Illustrated History of Tropical Diseases*. London, 1996: Wellcome Trust, pp. 356–69.

4 *Op cit.* See note 1 above (Boyd, 1966, 2004).

5 R Lainson, R Killick-Kendrick. Percy Cyril Claude Garnham CMG. *Biographical Memoirs of Fellows of the Royal Society*, Vol. 43. London, 1997: Royal Society, pp. 171–92; Anonymous. Professor P C C Garnham. *Medical News* 1965, April; P C C Garnham. Henry Edward Shortt. In: *Biographical Memoirs* (see above). 1988, 34: 715–51.

6 *Op cit.* See note 1 above.

7 N H Fairley. Schistosomiasis and some of its problems. *Trans R Soc Trop Med Hyg* 1951, 45: 279–303.

8 N H Fairley. The Hospital for Tropical Diseases (UCH), London. *Univ Coll Hosp Mag* 1952, 37: 114–18. See also P Manson-Bahr. *History of the School of Tropical Medicine in London (1899–1949)*. London, 1956: H K Lewis, pp. 64–7.

9 G C Cook. *John MacAlister's Other Vision: a history of the Fellowship of Postgraduate Medicine*. Oxford, 2005: Radcliffe Publishing, p. 178.

10 G C Cook. *Disease in the Merchant Navy: a history of the Seamen's Hospital Society*. Oxford, 2007: Radcliffe Publishing (in press). See also Anonymous. Minutes of the Royal Society of Tropical Medicine and Hygiene 1945: 18 January; G C Cook. Evolution: the art of survival. *Trans R Soc Trop Med Hyg* 1994, 88: 4–18.

11 G C Cook. *From the Greenwich Hulks to Old St Pancras: a history of tropical disease in London*. London, 1992: Athlone Press, p. 299; G C Cook. Tropical Medicine as a formal discipline is dead and should be buried. *Trans R Soc Trop Med Hyg* 1997, 91: 372–4.

16

Alexandre Yersin (1863–1943), and other contributers in solving the plague problem

Most of the discoveries leading to the creation of the formal discipline of tropical medicine involved protozoa and helminths (see Chapter 2), although in the case of yellow fever a virus was involved (see Chapter 6). However, the aetiological agent in plague is a bacterium, and research in the last decade of the nineteenth century finally clinched the solution to this age-old problem. The causative agent was discovered in 1894 by Alexandre Yersin (1863–1943), a pupil of Pasteur, working in Hong Kong.

HISTORY OF PLAGUE

Plague is, of course, a disease of great antiquity, as Scott has pointed out in his classical book – *A History of Tropical Medicine*.[1] It might have constituted the Plague of Athens (430 BC) and that of Justinian (AD 543), although proof in both cases is lacking.[2] The greatest pandemic in history, the Black Death (the cause of which is now disputed), centred on 1348; it began in the Crimea and ultimately reached England via southern Europe. The effect on the economic and agricultural structure of the country, associated with this massive mortality rate, was enormous.[3] Then came the Great Plague of London, in 1665. The disease had in fact been endemic in the capital since 1590, but in 1665 there were

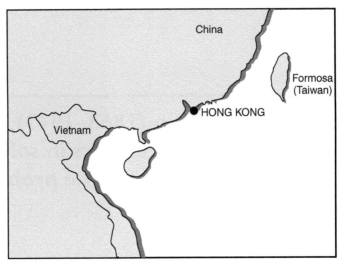

FIGURE 16.1 Map showing the location of the 1894 plague epidemic in Hong Kong, the site of isolation and identification of the causative agent.

over 68 000 deaths from this infection.[4] Apart from outbreaks in Marseilles and Egypt in 1720 and 1834 respectively, plague has remained a 'tropical' disease ever since. In 1894, plague was responsible for an estimated 100 000 deaths in Canton (China) (see Figure 16.1), and in that and the following year 1 300 000 died in India. However, when precisely this became an infection of warm climates is, as Scott has pointed out, impossible to say; in such an environment there is often an increased chance of rat fleas biting man, while dwellings there are on the whole more insanitary and overcrowded – a situation which is conducive to the proliferation of rats.

Discovery of the causative agent has been admirably described by Moote and Moote.[5] Identification of the plague bacillus, *Pasteurella* (now renamed *Yersinia*) *pestis* – which was in fact probably due to both Yersin and Kitasato, independently – did not immediately solve the problem of spread, though, because there was initially no recognition of an intermediate host.[6]

From early times until the last decades of the nineteenth century, there had been, as was the case with most other diseases, two major theories of causation; the 'miasma' and 'contagion' theories. Preventive measures were based on the assumption that both were correct!

THE CAUSATIVE AGENT

Alexandre Emile John Yersin (1863–1943; Figure 16.2), who is credited with the initial discovery, was a Swiss-born French bacteriologist. He had worked at the

FIGURE 16.2   Alexandre Yersin (1863–1943), the accepted discoverer of the causative organism of bubonic plague (reproduced courtesy of The Wellcome Library, London).

Pasteur Institute, under Pasteur, and the British Government requested that he determine the cause of bubonic plague in humans in the Hong Kong epidemic of 1894. Baron Shibasaburo Kitasato (1852–1931), a Japanese bacteriologist who had previously worked with Koch, also isolated what he considered was the causative agent of plague in the same year, independently of Yersin. Yersin's culture, the nature of which was later confirmed by rat experiments in Paris, consisted however solely of Gram-negative, non-motile organisms, unlike Kitasato's, which consisted of a mixed growth and was hence contaminated. Scott has emphasized that the name of Kitasato is often coupled with that of Yersin, although he probably 'mistook some contaminent for the causative organism, which was neither Gram-positive nor motile', and made this claim about three weeks before Yersin's correct (uncontaminated) identification.

Yersin developed an antiserum for plague, which proved valuable in 1898 in prevention of a further outbreak in Hong Kong; he also established two Pasteur Institutes in China and, incidentally, introduced the rubber tree into China.[7]

INVOLVEMENT OF RATS – THE INTERMEDIATE HOST

Although the association between rats and plague had been noted both in ancient times and in China in 1792, 1834 and 1894, and latterly the plague bacterium had been isolated from dead rats, the rat–human cycle was largely discounted because rats could not always be found in a 'plague locality'.

Scott has quoted from the translated poem of a Chinese writer, Shih Taonan (1765–92), who in fact died of plague:[8]

> Dead rats in the east,
>> Dead rats in the west! ...
> Few days following death of the rats,
>> Men pass away like falling walls!
> Deaths in the day are numberless,
>> The hazy sun is covered by sombre clouds.
> While three men are walking together
>> Two drop dead within ten steps!
> People die in the night
>> Nobody dares weep over the dead!
> The coming of the devil of plague
>> Suddenly makes the lamp dim,
> Then it is blown out,
>> Leaving man, ghost and corpse in the dark room.
> The land is filled with human bones,
>> There in the fields are crops,
> To be reaped by none;
>> And the officials collect no tax!!

In a maritime context, an association between rats and plague had been noted for many years; this led to destruction of rats in ships, and the introduction of strict quarantine. As Scott has written: 'In the days of rat-infested wooden vessels the danger of transmission of plague by sea or inland navigation was great'.[9]

Paul Lewis Simond (1858–1947), working in Bombay, first propounded the 'rat flea' theory,[10] but this was not confirmed, by A W Bacot (1866–1922) and C Martin (1866–1915), until 1914.

The black rat (*Rattus rattus*) is a far more effective carrier of the rat flea (*Xenopsyla cheopis*) than is the brown one (*Rattus norvegicus*); the former rodent was dominant in western Europe, including the British Isles, between the early eighteenth and nineteenth centuries. A change in the rat population from black to brown therefore largely explains the disappearance of plague from Europe in the late nineteenth and early twentieth centuries. However, since 1910 *R. rattus* has gained ground, and plague prevention is again largely dependent on strict quarantine at all ports of entry – including airports. The history of quarantine is complex, and has been admirably reviewed by Scott.[11]

Other vectors

Certain wild rodents, such as marmots, can also act as hosts of the plague bacillus, and they convey a frequently fatal form of the disease – the pneumonic rather than the bubonic form, which is responsible for numerous deaths in some isolated communities of Asia, South Africa, California, etc.[12]

EPILOGUE

Yersinia pestis infection is essentially a disease of rodents, and in the bubonic form man becomes infected accidentally. In the more unusual pneumonic form, man can pass the infection to another individual without the intervention of the flea. However, as Rogers (see Chapter 13) has succinctly pointed out, factors such as climate, temperature and humidity all play a part in the epidemiology of what is now a *tropical* infection.[13]

NOTES

1 G C Cook. Scott, Sir (Henry) Harold (1874–1956). In: H C G Matthew, B Harrison (eds), *Oxford Dictionary of National Biography*, Vol. 49. Oxford, 2004: Oxford University Press, pp. 386–7; H H Scott. *A History of Tropical Medicine*, two volumes. London, 1939: Arnold, pp. 702–67; L B Hirst. *The Conquest of Plague: a study of the evolution of epidemiology.* Oxford, 1953: Clarendon Press, p. 478; C Singer, E A Underwood. *A Short History of Medicine*, 2nd edn. Oxford, 1962: Clarendon Press, pp. 488–92.

2 R S Bray. Plague. In: F E G Cox (ed.), *The Wellcome Trust Illustrated History of Tropical Diseases.* London, 1996: Wellcome Trust, pp. 40–49; E Marriott. *The Plague Race: a tale of fear, science and heroism.* London, 2002: Picador, p. 275; W Orent. *Plague: mysterious past and terrifying future of the world's most dangerous disease.* New York, 2004: Free Press, p. 276.

3 J Nohl. *The Black Death: a chronicle of the Plague.* London, 1926: George Allen & Unwin, p. 284; D Williman (ed.). *The Black Death: the impact of the fourteenth-century plague.* New York, 1982: Medieval & Renaissance Texts and Studies, p. 159; G Twigg. *The Black Death: a biological reappraisal.* London, 1984: Batsford, p. 254; N F Cantor. *In the Wake of the Plague: the Black Death and the World it made.* New York, 2001: Free Press, p. 245; W Naphy, A Spicer. *Plague: Black Death and pestilence in Europe.* Stroud, 2004: Tempus, p. 222; J Kelly. *The Great Mortality: an intimate history of the Black Death, the most devastating plague of all time.* New York, 2005: Harper Collins, p. 364. See also P Ziegler. *The Black Death.* London, 1969: Collins, p. 319; Sale Bruxelles. Lost documents shed light on Black Death. *Times*, London, 2007, 1 June: 25.

4 W G Bell. *The Great Plague of London.* London, 1994: Bracken Books, p. 374; S Porter. *The Great Plague.* Stroud, 1999: Sutton Publishing, p. 213; A L Moote, D C Moote. *The Great Plague: the story of London's most deadly year.* Baltimore, 2004: Johns Hopkins University Press, p. 357.

5 *Op cit.* See note 4 above (Moote and Moote).

6 *Op cit.* See note 1 above (Scott; Singer, Underwood).

7 H H Mollaret, J Brossollet. *Alexandre Yersin: ou le vainquer de la peste.* Paris, 1985: Fayard, p. 317. See also Anonymous. *Institut Pasteur: MCMXXXVIII.* Paris, 1939: J Dumoulin, p. 69.

8 *Op cit.* See note 1 above (Scott, p. 731).

9 *Op cit.* See note 1 above (Scott, p. 750).

10 P L Simond had also previously worked at the Pasteur Institute, Paris.

11 *Op cit.* See note 1 above (Scott pp. 751–7).

12 *Ibid.* See also: *Op cit.* See note 1 above [Singer, Underwood]; M D Smith. Plague. In: *Manson's Tropical Diseases*, 21st edn. London, 2003: W B Saunders, pp. 1125–31.

13 *Ibid.*

17

Andrew Balfour (1873–1931): pioneer of preventive medicine in the tropics and first Director of the London School of Hygiene and Tropical Medicine

Andrew Balfour (1873–1931; Figure 17.1) was born in Edinburgh on 21 March 1873. He was educated at George Watson's College and the University of Edinburgh, where he graduated MB CM in 1894 and MD in 1898. After a short period in private practice with his father, he entered Gonville and Caius College, Cambridge, as an advanced student, and obtained the DPH in 1897. He also graduated BSc in public health at Edinburgh in 1900. He worked on typhoid at Cambridge and in Pretoria, where he contracted the infection. During the South African War, he served as a civil surgeon in the Transvaal.

In 1901, contact with Manson (see Chapter 3) initiated his interest in the *tropical medicine* specialty. In 1902 he became director of the Wellcome Tropical Research Laboratories at Khartoum, Medical Officer of that city, and Chief Scientist for The Sudan, member (with Lords Cromer and Kitchener, and Sir Reginald Wingate) of the Sleeping Sickness Commission, and founder of the floating laboratory (donated by H S Wellcome to the Sudan Government) on the Nile and White Nile (see Figure 17.2). On the floating laboratory was his colleague C M Wenyon, a protozoologist from Liverpool. Overall, he spent eleven years in Khartoum. He strongly advocated care and health of the indigenous population of the Sudan as essential features of Imperial rule. Later, in the West Indies, he was also a member of the Colonial Office Commission on Health in

FIGURE 17.1   Sir Andrew Balfour (1873–1931) (reproduced courtesy of The Wellcome Library, London).

the Colonies. He presided at the Commission in Mesopotamia, and was scientific adviser to the Commission on East Africa.[1]

In 1913, Balfour returned to England and served in the RAMC as a Lieutenant-Colonel. In 1918 he became Director of the Wellcome Research Laboratories and founder of the Museum of Medical Sciences; he had already founded the Tropical Diseases Bureau in 1912.

During the Great War, Balfour saw service in France, Madras, Salonica, Egypt and Mesopotamia; he later served in Tanganyika Territory (now Tanzania) and Uganda. He also visited Palestine at the request of General Allenby, in order to investigate anti-malarial strategies there.

In 1923, Balfour became the first (and only) Director of the London School of Hygiene and Tropical Medicine (LSHTM).

Andrew Balfour's contributions to medicine in the tropics, most of a preventive nature, were numerous. He virtually cleared Khartoum of mosquitoes,

FIGURE 17.2 Balfour's floating laboratory on the river Nile (reproduced courtesy of The Wellcome Library, London).

discovered the life-cycle of ticks causing spirochaetosis, and, during his work in Sudan on protozoa, described a leishmanoid skin disease. Furthermore, he made a prediction – which was later confirmed – that the Red Howler monkey in Trinidad constituted an animal reservoir of yellow fever (see Chapter 6).

Balfour had formerly excelled at boxing and rugby football, representing Scotland at the latter, and was later appointed to the governing committee of the Rugby Union. He was a good lecturer and writer, in his early years producing several novels. Balfour was also a talented amateur actor. With C J Lewis, he edited *Public Health and Preventive Medicine* (1902), the *War Office's Memoranda on Medical Diseases in the Tropical and Sub-tropical War Areas* (third edition 1919) – formerly entitled *Memoranda on some Medical Diseases in the Mediterranean War Area* (1916), *War Against Tropical Diseases* (1920), *Reports to the Health Committee of the League of Nations on Tuberculosis and Sleeping Sickness in Equatorial Africa* (1933) and, with H H Scott, *Health Problems of the Empire* (1924).

This otherwise warm-hearted man was apparently of cyclo-thymic temperament, eventually committing suicide while undergoing management of depression. On 30 January 1931, while receiving treatment for a 'nervous breakdown' at the Cassel Hospital at Penshurst, Kent, his body was discovered frozen stiff in the hospital grounds, with a sash-cord around the neck. He had suffered from insomnia and was concerned that he might be compelled to resign his post at the LSHTM. At an inquest, death was considered to have been brought about by asphyxiation, and a verdict that he 'had strangled himself while temporarily of unsound mind' was returned.[2]

In 1912 Balfour was appointed CMG, in 1918 CB, and in 1930 KCMG. He received several honorary degrees, and the Mary Kingsley Medal in 1920.

Balfour was thus a major twentieth-century pioneer of public health (hygiene) in tropical countries. Before him was Ross (see Chapter 5) who, following his Nobel Prize-winning work, was a great enthusiast of preventive rather than curative medicine. Another example was Sir Malcolm Watson (1873–1955), who carried out preventive strategies in Malaya and later in Northern Rhodesia (now Zambia). Balfour also guided the formative years of the premier school of public health in Britain, the LSHTM (see below), from its initiation.[3]

GROWTH OF PUBLIC HEALTH (HYGIENE)

As Acheson and Poole have pointed out, 'the idea that health could be promoted, and indeed that medicine could be practised by the management and treatment of *populations* [my italics], gradually gathered [momentum] during the nineteenth century'. By 1870, it was widely appreciated, with both the British Medical Association and General Medical Council being of one mind, that appropriate instruction in 'State medicine'[4] should be introduced.

Long after the introduction of institutes in several European countries, the Rockefeller Foundation – through Wickliffe Rose, with R J Leiper (see Chapter 10) as a 'go-between' – decided, with support from the British Government (in the person of Sir George Newman), to erect a School of Public Health or 'State medicine' in London, the heart of the British Empire. Immediately before the outbreak of the Great War (1914–18), the Rockefeller Foundation had been grappling with its 'hookworm campaign', the object of which was to rid the southern states of the debilitating effects of ankylostomiasis (see Chapter 10). The introduction, soon after cessation of hostilities, of a scheme at the heart of the British Empire of great potential beneficence, was deemed to be a highly desirable objective; this disease should, they considered, be eliminated from the British Empire. Establishment of a Public Health School had been recommended in the Athlone Report (1921),[5] which had been directed at the Minister of Health, Sir Alfred Mond. Leishman (see Chapter 12), now Medical Director-General in the British Army, emphasized that preventive medicine in the tropics differed significantly from that at home; however, this distinction between domestic and foreign objectives seems to have been very largely ignored.

Late in 1905, the Lister Institute of Preventive Medicine (which it was intended would be the equivalent of the Institute Pasteur in Paris, and therefore concentrate on bacteriology)[6] had joined the London School of Tropical Medicine (LSTM) as a new member of London University's Faculty of Medicine. There was a continuing debate about where the University Chair of Bacteriology should be located – at the LSTM or the Lister! Until Wickliffe Rose, with Leiper's assistance, became involved, it had been assumed that a Centre of Public Health (or Hygiene) would be created at University College, London. Manson, who was by this time nearing the end of his life, would have clearly preferred the LSTM to have remained separate from this new Institute, but this was not to be!

If Wickliffe Rose had not been able to persuade the British Government to found this new School of Hygiene, 'State medicine' would have been established at University College, London. Foundation of the LSHTM was *not* supported by the SHS, although it continued sponsoring the HTD after 1924. The LSHTM was 'under the sway of the Interim Executive Committee [until completed]'. It became a reality in 1924, and the Foundation Stone was laid on 10 July 1926 (see Figure 17.3). According to Manson-Bahr, 'a scrutiny of [the recommendations of the Athlone Report] reveals hardly any reference to Tropical Medicine'; the report was principally concerned with the 'setting up [of] a Post-graduate Medical School'. The 'fusion of tropical medicine with hygiene [wrote Manson-Bahr] was a new idea and to some it signified a first move away from Manson's original conception [and this was inevitably linked to] a decline of tropical medicine as an *independent* [my italics] subject'.

ORIGIN(S) OF THE LSHTM

Manson-Bahr, and Acheson and Poole, have provided valuable accounts of the early days of the LSHTM.[7] The creation of this Public Health School, a product of

FIGURE 17.3   Foundation Stone of the London School of Hygiene and Tropical Medicine, which was laid by the Minister of Health, the Rt Hon. Neville Chamberlain, in 1926.

growing altruism in the USA, was in effect a triumph for the American Wickliffe Rose, who was able to persuade the British Government, in the days of Empire, to establish this foundation with USA funding.

Balfour's greatest achievement

The Minister of Health, Joynson Hicks, acting on a recommendation of the Interim Executive Committee, appointed Balfour as Director of the new School on 6 November 1923; this was the date, wrote Manson-Bahr, on which the LSHTM was in fact founded. Balfour was very largely personally responsible for shaping this new School; he toured the Institutes of Health and the Rockefeller Institute in the USA and, according to Manson-Bahr, was a frequent visitor to the HTD at Endsleigh Gardens. The school was officially opened by the Prince of Wales (later King Edward VIII) on 18 July 1929. Wilson Jameson, Balfour's successor as Dean (*not* Director) of the LSHTM and a future Principal Medical Officer, became Director of the Division of Public Health. This School (of the University of London) celebrated its centenary in 1999; however, although this marked 100 years since the foundation of Manson's School at the Albert Dock Hospital, the LSHTM was a mere 70 years old.[8]

Balfour, whose main interest was in prevention rather than cure, was thus Director of the LSHTM for only some eighteen months following the formal opening ceremony – he died on 30 January 1931 – although for no less than nine years this venture had absorbed most of his time.

NOTES

1   P F D'Arcy. *Laboratory on the Nile: a history of the Wellcome Tropical Research Laboratories.* London, 1999: Pharmaceutical Products Press, p. 281; Anonymous. Dr C M Wenyon: authority on tropical medicine. *Times, Lond* 1948, 26 October.

2   Anonymous. *Times, Lond* 1931, 7 February: 9; G C Cook. Fatal yellow fever contracted at the Hospital for Tropical Diseases, London, UK, in 1930. *Trans R Soc Trop Med Hyg* 1994, 88: 712–13.

3   Anonymous. *Times, Lond* 1931: 2 February; C M Wenyon. Sir Andrew Balfour. *Nature, Lond* 1931, 127: 279–81; Anonymous. Sir Andrew Balfour. *Lancet* 1931, i: 325–7; P H Manson-Bahr. Sir Andrew Balfour. *Br Med J* 1931, i: 245–6; C M Wenyon. Sir Andrew Balfour. *Trans R Soc Trop Med Hyg* 1931, 24: 655–9; Anonymous. *Munk's Roll*, Vol. 5. London, Royal College of Physicians, pp. 19–29; P Manson-Bahr. *History of the School of Tropical Medicine in London (1899–1949).* London, 1956: H K Lewis, pp. 167–73, also pp. 216–18; A S MacNalty, M E Gibson. Balfour, Sir Andrew (1873–1931). In: H C G Matthew, B Harrison (eds) *Oxford Dictionary of National Biography*, vol. 3. Oxford, 2004: Oxford University Press, pp. 493–4; M Watson. *African Highway.* London, 1953: John Murray, p. 294; A Balfour, why hygiene pays. *Br Med J* 1926, i: 929–32.

4   Alternative names were 'preventive medicine' and 'public health'. In 1895, the General Medical Council indicated its preferred opinion that the DPH (Diploma in Public Health) be used rather than diplomas in State Medicine, Sanitary Science, or Hygiene.

5   G C Cook. *John MacAlister's Other Vision: a history of the Fellowship of Postgraduate Medicine.* Oxford, 2005: Radcliffe Publishing, p. 178.

6 H Chick, M Hume, M Macfarlane. *War on disease: a history of the Lister Institute*. London, 1971: Andre Deutsch, p. 251.

7 *Op cit*. See note 3 above (Manson-Bahr), pp. 64–71; R Acheson, P Poole. The London School of Hygiene and Tropical Medicine: a child of many parents. *Med Hist* 1991, 35: 385–408.

8 Anonymous. Laying of the Foundation Stone of the London School of Hygiene and Tropical Medicine. *Br Med J* 1926, ii: 75–8; Anonymous. The London School of Hygiene and Tropical Medicine the opening of the new building. *Lancet* 1929, ii: 175–6; A May. *London School of Hygiene & Tropical Medicine 1899–1999*. London, 1999: London School of Hygiene and Tropical Medicine, p. 40. See also *Op cit*. See note 4 above.

18

Some less well-documented pioneers

The early days of the formal discipline were dominated by 'prima donnas' such as Manson, Ross, Bruce and Leishman. However, there were numerous 'lesser mortals', some of whom are given coverage in this chapter. All of their major contributions – to virology, bacteriology, protozoology or helminthology – were made *after* the 'germ theory' had been widely accepted; they thus added significantly to the formal discipline upon which most of this book has focused.

HIDEYO NOGUCHI

Hideyo Noguchi (1876–1928; Figure 18.1) was born at Inawashiro, Yama, Fakushima, Japan; as a child he suffered from a serious injury which resulted in deformity and partial paralysis of his left hand. He entered the University of Tokyo at seventeen years of age, and graduated in medicine in 1897. From 1898 until 1900, he was assistant to Kitasato at the Institute for Infectious Disease at Tokyo and also lectured on bacteriology at the Dental Institute. In 1901, he began work as an assistant pathologist with Simon Flexner at the University of Pennsylvania, USA, and in 1903 he became head of the research department of the Carnegie Institute, Washington. Following a brief period in Denmark,

PROF. HIDEYO NOGUCHI.

FIGURE 18.1 Hideyo Noguchi (1876–1928) (reproduced courtesy of The Wellcome Library, London).

he obtained a position in the recently opened laboratories of the Rockefeller Institute, where he remained until his death.

Noguchi's initial research was on toxins, antitoxins, agglutinins, haemolysins and venoms (upon which the Carnegie Institute published his monograph in 1909). Turning his attention to spirochaetes, he demonstrated their presence in the cerebral cortex of patients who had died of General Paralysis of the Insane (GPI), and in the spinal cord of tabetics. He then worked on trachoma, Oroya fever and *Verruga peruviana*, and originated the name *Bartonella* sp. He later studied other infections, including poliomyelitis, rabies, hog cholera, herpes, and Rocky Mountain spotted fever.

In 1918, Noguchi travelled to South America (Peru, Brazil and Mexico) in order to study yellow fever, which was felt by many at that time to be a spirochaetal disease (see Chapter 6). He coined the term *Leptospira icteroides* (later recognized to be the cause of Weil's disease – see below), which he temporarily considered to be the cause of yellow fever; however, workers in West Africa

GERHARD HENRIK ARMAUER HANSEN (1841–1912).

FIGURE 18.2   Gerhard Hansen (1841–1912), discoverer of the causative agent of leprosy (reproduced courtesy of The Wellcome Library, London).

failed to detect this organism in cases of the disease. After a considerable delay he travelled to Accra, Gold Coast (now Ghana) in late 1927, and he too failed to detect *L. icteroides* there. Following a brief visit to Lagos, Nigeria, he returned to Accra, where he died of yellow fever on 21 May 1928. The true causative agent of yellow fever was ultimately unravelled (see Chapter 6).[1]

GERHARD HENRIK ARMAUER HANSEN

Gerhard Henrik Armauer Hansen (1841–1912; Figure 18.2) was the discoverer of the causative agent of leprosy. This is of course an ancient disease, which has only relatively recently become 'tropical'. Numerous theories of its causation had been suggested, many of them genetic.

Hansen was born in Bergen, Norway, qualifying in medicine in 1866. After about two years at Bergen's Leprosy Hospital he underwent a period of study in Vienna, where he became familiar with the 'germ theory' of disease (see Chapter 2). Returning to Bergen, Hansen was by now convinced that leprosy was a specific communicable disease. With a grant from the Norwegian Government, Hansen first carried out an epidemiological survey of the disease in Norway and concluded that it was certainly *not* genetically determined.

On 28 February 1873, he viewed an unstained preparation of nasal biopsies from a patient suffering from the disease; large numbers of rod-shaped bodies were present. Later that year he submitted this finding to a medical society at Christiana, and the following year, 1874, H V Carter of the Bombay Army visited Bergen and visualized for himself Hansen's finding. Later that year, Hansen recorded the discovery in a (British) White Paper.

Later, Hansen inoculated leprous material into the cornea of a female patient; as a sequel, a public court banned him from practising in hospitals in Norway for the remainder of his life. He did, however, continue as an adviser on leprosy to the Government, and also continued his researches.[2]

ADOLF WEIL

Adolf Weil (1848–1916; Figure 18.3) was Professor of Medicine at Heidelberg University. In 1886 he published four cases of an acute febrile illness associated with jaundice, severe neurological symptoms, hepato-splenomegaly and renal involvement; the course was short and recovery rapid. In three cases, an afebrile phase (of one to seven days) was followed by a recurrence of fever (for five or six days). The term 'Weil's disease' was coined by Goldsmith.

For almost 30 years, the diagnosis of Weil's disease was of necessity a clinical one. In 1907, Stimson recorded an organism resembling a spirochaete (*Spirocheata interrogans*) in the renal tubules of a patient diagnosed on clinical grounds as suffering from yellow fever (see Chapter 6). In the autumn of 1914 (published in February 1915) Ryokichi Inada of the Imperial University, Kyusha, Japan, found spirochaetes – *Spirochaeta ictero-haemorrhagica japonica* – in the liver of a guinea-pig inoculated with blood from a patient suffering from Weil's disease. The original work was published in Japanese, and it was not until March and August 1916 that the discovery appeared in the English and German medical press, respectively – as *Spirochaeta ictero-haemorrhagiae*. German workers also demonstrated the organism in guinea-pigs inoculated with the blood of patients suffering from Weil's disease, but not until October 1915.

A fellow Japanese worker, Noguchi (see above), studied Inada's *S ictero-haemorhagiae*. Adrian Stokes (1887–1927) also investigated cases in Flanders (this was a common disease in the trenches of the Great War), and they came to the conclusion that both were the same. Noguchi then created a new

FIGURE 18.3 Adolf Weil (1848–1916), the originator of an eponymous disease (reproduced courtesy of The Wellcome Library, London).

genus – *Leptospira*. Approximately 2000 'serovars', only some of which are pathogenic, have now been differentiated.

Confusion arose in 1919, when Noguchi recorded an organism which he called *L. icteroides* in guinea pigs inoculated with blood from patients with supposed yellow fever. (In retrospect, many cases diagnosed as suffering from yellow fever must in fact have suffered from Weil's disease.) Noguchi considered that this organism was its cause! However, workers at the Rockefeller laboratory at Yaba, near Lagos, Nigeria (under the aegis of the West African Yellow Fever Commission) failed to detect Noguchi's organism in the blood of patients suffering there from yellow fever.

Stokes, then Professor of Pathology at Guy's Hospital, London, joined the researchers at Yaba in 1927, and duly developed yellow fever; he insisted that his blood be inoculated into monkeys, and furthermore that mosquitoes be allowed to bite him. From these sources the yellow fever virus was identified, and Noguchi's organism was discredited as the cause of this disease. Stokes died of this infection on 19 September 1927. In May 1928, Noguchi, who had travelled to West Africa with the expressed purpose of confirming that his organism was indeed causative in yellow fever, also died of the infection in Accra, Gold Coast (now Ghana).[3]

OSWALDO GONÇALES CRUZ

Oswaldo Gonçales Cruz (1872–1917; Figure 18.4) was born at São Luis de Parahitinga, São Paulo, Brazil, but the year after his birth his parents moved to Rio de Janeiro, where he graduated in medicine from the University in 1892. He then entered the National Institute of Hygiene, which had then been founded by Rocha Faria, to study bacteriology, and four years later he began a study of toxicology at the Pasteur Institute, Paris.

Cruz was so impressed by the preventive work of the American Yellow Fever Commission (which advocated the extermination of mosquitoes by both attacking

FIGURE 18.4 Oswaldo Cruz (1872–1917), who performed public health researches in southern America (reproduced courtesy of The Wellcome Library, London).

their breeding sites and preventing contact between adult *Aëdes* and infected patients) in reforming the sanitation of such paces as Havana and Santiago (see Chapter 6) that he decided to apply these principles to Rio de Janeiro, where yellow fever was rife. In 1903, Cruz was appointed Director-General of the Office of Public Health and, by applying these principles, achieved a staggering result; the incidence not only of yellow fever but also of plague diminished dramatically. He later applied these principles to the whole of Brazil.

In 1908 Cruz became Head of the Institute for Tropical Diseases at Rio, and remained its Director until his death. One of his pupils was Carlos Chagas (see below), who discovered the causative agent of South American trypanosomiasis and named the organism *Trypanosoma cruzi* in honour of Cruz. Cruz had thus assumed important roles in the unravelling of both the yellow fever and the South American trypanosomiasis sagas.[4]

CARLOS CHAGAS

Carlos Chagas (1879–1934; Figure 18.5) studied South American trypanosomiasis, now named Chagas' disease, the history of which has been outlined by Miles.[5] Chagas demonstrated both the aetiological agent and its transmission by the reduviid (triatomine) bug; a suggestion that Charles Darwin (1809–82) suffered from this infection has now been discredited.

Chagas made his important discoveries between 1907 and 1909. After graduation from the University of Rio de Janeiro, he was appointed in 1902 to the Institute of which Cruz (see above) was Director. Chagas was greatly assisted in his discoveries by Cruz, who was able to demonstrate that inoculation of the causative agent to monkeys, and several other laboratory animals, proved pathogenic.[6]

SIR PHILIP HENRY MANSON-BAHR

Sir Philip Henry Manson-Bahr (1881–1966; Figure 18.6) was born in Liverpool and educated at Rugby School, followed by Trinity College, Cambridge, and the (Royal) London Hospital. His early interests were in ornithology and zoology. Bahr married Manson's (see Chapter 3) daughter in 1909; she died in 1949.

Bahr's first overseas 'posting' was as head of the Stanley Research Expedition to Fiji, where he studied dysentery (he both isolated *Shigella shiga* from housefly and showed that the macrophages were easily mistaken for vegetative *E histolytica*), and lymphatic filariasis. From 1912 until 1913 he travelled thousands of miles by bicycle in Ceylon (now Sri Lanka), studying 'sprue', filariasis and malaria; his researches on the former were published as a monograph in 1915.

During the Great War (1914–18) Manson-Bahr served in Egypt, Palestine and the Dardenelles, during which time he dealt with dysentery, malaria, cholera and pellagra. Following the war, he was appointed to the Albert Dock Hospital

Carlos **Chagas**

FIGURE 18.5 Carlos Chagas (1879–1934), discoverer of the causative agent of South American trypanosomiasis (reproduced courtesy of The Wellcome Library, London).

and, later, the Hospital for Tropical Diseases. He also became a lecturer at the London School of Hygiene and Tropical Medicine (see Chapter 17), and from 1937 until 1947 was Director of its Department of Clinical Tropical Medicine. From 1927 until 1947 Manson-Bahr was Consultant Physician to the Colonial Office and the Crown Agents. In addition to numerous papers on various aspects of tropical medicine, he edited *Manson's Tropical Diseases* from 1921 (seventh edition) until 1960 (fifteenth edition).

In his Presidential Address to the Royal Society of Tropical Medicine and Hygiene in 1947, Manson-Bahr concentrated on the *clinical* aspects of tropical medicine as practised in London – to some degree, the principal object of this Society. His experience was based on 'an analysis of some 5,600 hospital records as well as probably almost as many again seen in consultation'. The Address consisted largely of elucidation of physical signs, punctuated by numerous anecdotes. He also went to great lengths to explain, quite correctly, that in the practice of the

FIGURE 18.6 Sir Philip Manson-Bahr (1881–1966) (reproduced courtesy of The Wellcome Library, London).

discipline it is important to include non-tropical conditions in the differential diagnostic list; the 'tropical specialist' has thus (he suggested) to be a general physician with a special knowledge of the diseases peculiar to warm climates. Of particular note is Manson-Bahr's contention in this address (*not* supported by statistics) that sprue had declined dramatically in prevalence during the previous ten years![7]

In later life, Bahr changed his name by deed-poll to Manson-Bahr; apparently this was Manson's wish. He later received many honours, including the Bernhardt Nocht Medal (of the Tropeninstitut, Hamburg), the Mary Kingsley Medal (Liverpool School of Tropical Medicine), and the Brumpt Prize (Paris). Manson-Bahr was also an accomplished water-colourist and ornithologist.[8]

CONCLUSION

This chapter has concentrated on six less well-documented pioneers than the 'prima donnas' whose careers have been outlined in earlier chapters; all of their contributions came historically *after* general acceptance of the 'germ theory' of disease, and thus fell after the formal discipline had been established. However, many of their research findings also have application to the older discipline, 'medicine in the tropics', which will continue until *Homo sapiens* departs from this planet.

NOTES

1 I R Plesset. *Noguchi and his Patrons.* London, 1912: Rutherford, p. 314; G Eckstein. *Noguchi.* London, 1931: Harper and Brothers, p. 419; H H Scott. *A History of Tropical Medicine.* London, 1939: Arnold, pp. 1077–80; A Kita. *Dr Noguchi's Journey: a life of medical search and discovery.* Tokyo, 2005: Kodansha International, p. 252.

2 F B Watt. *Les mémoires de Hansen.* Québec, 1976: Les presses de l'université du Québec, p. 130; A C McDougall. Leprosy. In: F E G Cox (ed.). *The Wellcome Trust Illustrated History of Tropical Diseases.* London, 1996: Wellcome Trust, pp. 60–71.

3 J D Everard. Leptospirosis. In: F E G Cox (ed.). *The Wellcome Trust Illustrated History of Tropical Diseases.* London, 1996: Wellcome Trust, pp. 110–19. See also K Duncan. Climate and the decline of leprosy in Britain. *Proc R Coll Phys Edin* 1994, 24: 114–20.

4 *Op cit.* See note 1 above (Scott), pp. 1024–6.

5 M A Miles. *New World Trypanosomiasis.* In: F E G Cox (ed.), *The Wellcome Trust Illustrated History of Tropical Diseases.* London, 1996: Wellcome Trust, pp. 192–205.

6 F Guerra. American trypanosomiasis: an historical and a human lesson. *J Am Trop Med Hyg* 1970, 72: 83–118; R Lewinsohn. The discovery of *Trypanosoma cruzi* and of American trypanosomiasis (footnote to the history of Chagas's disease). *Trans R Soc Trop Med Hyg* 1979, 73: 513–23; M A Miles. *New World Trypanosomiasis.* In: F E G Cox (ed.), *The Wellcome Trust Illustrated History of Tropical Diseases.* London, 1996: Wellcome Trust, pp. 192–205; M A Miles. The discovery of Chagas: progress and prejudice. *Infect Dis Clin North Am* 2004, 18: 247–60.

7 P Manson-Bahr. The practice of tropical medicine in London. *Trans R Soc Trop Med Hyg* 1947, 41: 269–94. See also M L Lim, M R Wallace. Infectious diarrhoea in history. *Infect Dis Clin North Am* 2004, 18: 261–74.

8 Anonymous. Sir P Manson-Bahr: leading tropical medicine consultant. *Times, Lond* 1966, 21 November; Anonymous. Philip Henry Manson-Bahr. *Lancet* 1966, ii: 1198–9; Anonymous. Sir Philip Manson-Bahr. *Br Med J* 1966, ii: 1332; Anonymous. *Ibis* 1967, 109: 447–8; A W Woodruff. Manson-Bahr, Sir Philip Henry. *Munk's Roll*, Vol. 6. London: Royal College of Physicians, pp. 328–39. See also: P H Bahr. *A Report on Researches on Sprue in Ceylon 1912–1914.* Cambridge, 1915: Cambridge University Press; G C Cook. Tropical sprue. In: F E G Cox (ed.), *The Wellcome Trust Illustrated History of Tropical Diseases.* London, 1996: Wellcome Trust, pp. 356–69.

19

'Back-room' and lay pioneers of the specialty

The bulk of this book has focused on the principal investigators – most of them protozoologists or helminthologists – of the newly-established discipline. This chapter deals with the 'back-room' contributors, most of whom were dominant figures in the Liverpool and London Schools. Similar contributors to other nascent institutions are not included because they were not, by definition, pioneers. As Low has recorded, other European colonial powers initiated their own schools in the wake of England's experience – Hamburg in Germany, Antwerp in Belgium, Paris in France, and Lisbon in Portugal. However, the USA did not develop a centre of excellence which was entirely dedicated to the study of disease in the tropics, although departments dealing with tropical diseases were established at Tulane (Louisiana), Johns Hopkins, Harvard, the National Institutes of Health, the Walter Reed Army Institute of Research, and the Rockefeller Institute.

In London especially – the centre of the formal discipline – there was a handful of back-room staff who were essential for the newly-established discipline to get off the ground. None was more important than the Superintendents (later Directors) of the London School of Tropical Medicine (LSTM) and, later, the Hospital for Tropical Diseases. Although perfectly adequate, the first two Deans of the LSTM – Francis Lovell (1844–1916) and Havelock Charles

(1858–1934) – seem to have been nothing more than competent administrators and fund-raisers, and far less inspirational than their equivalents in Liverpool. Teaching was, of course, of paramount importance, and it should be recalled that the new discipline was founded largely as a result of Manson's lectures on the specialty, *tropical medicine*. Furthermore, the Superintendent had often to 'hold the fort' while the Heads of Departments were abroad carrying out their pioneering researches.[1]

THE LIVERPOOL SCHOOL OF TROPICAL MEDICINE

Sir Rubert William Boyce FRS (1863–1911)

Sir Rubert William Boyce's (1863–1911; Figure 19.1) name should be linked to that of Sir Alfred Jones (see Chapter 20); both took a prominent role in the foundation of the Liverpool School of Tropical Medicine.[2] Boyce was the first Dean

FIGURE 19.1 (Sir) Rubert Boyce (1863–1911), first Dean of the Liverpool School of Tropical Medicine (reproduced courtesy of The Wellcome Library, London).

of the School, and occupied that position for seven years. During this period the School became firmly established, and also organized seventeen expeditions to various tropical countries – far more, in fact, than the London School. No less than 32 expeditions are recorded in the School's first history, which ends in 1920.[3] The Liverpool School's inaugural dinner, with Lord Lister (1827–1912) present, was held on 22 April 1899.

Boyce was born in London and educated in Rugby, Paris, and University College Hospital, London; He became a pathologist, trained originally at University College, London and, later, Liverpool (see below). Following graduation (MB, BS) in 1889 he worked in both Heidelberg and Paris, and under Victor Horsley at University College, London.

In 1894, Boyce was appointed first occupant of the George Holt Chair of Pathology at University College, Liverpool (forerunner of the University of Liverpool), then a component of Victoria University, Manchester. Whilst there, he organized a School of Hygiene. In 1898, he was also appointed bacteriologist to the City of Liverpool. Following this, he took a major role in the establishment of the University of Liverpool, which opened in 1902.

Boyce became 'particularly interested in the relationship between Britain and its colonies', and in April 1899 founded the Liverpool School, with the financial backing of Alfred Jones (see Chapter 20). He was appointed first Dean, with Ross (see Chapter 5) the Director. It was Boyce who organized the first expeditions to tropical countries; in 1905 he himself travelled to New Orleans and British Honduras to study epidemics of yellow fever (see Chapter 6).

In 1902 Boyce was elected FRS for his research on the nervous system, and four years later he was knighted. In 1909, three years after suffering a stroke, he visited the West Indies; he travelled there, and also to West Africa, essentially to advise governments on prevention of yellow fever at these two territories. He was also a member of the Colonial Office's West African Advisory Board, and served on Royal Commissions devoted to sewage disposal and tuberculosis. He wrote two books, in 1909 and 1910 respectively, on various aspects of health in the tropics for a lay readership.

Above all, Boyce was a highly skilled fundraiser, and early success of the Liverpool School was very largely due to his endeavours.[4]

THE LONDON SCHOOL OF TROPICAL MEDICINE

David Charles Rees (1868–1917)

David Charles Rees was the first Superintendent and medical tutor of the London School at the Albert Dock Hospital. He was already on the staff of the ADH, and was identified by Manson (see Chapter 3) as a suitable candidate for the post of Superintendent of the newly established School. Rees was largely responsible for the School's curriculum and syllabus, which apparently remained unaltered for many years.

Rees was born in London and received his medical education at Charing Cross Hospital, qualifying MRCS, LRCP in 1895. He had previously served in the West African Frontier Force campaign of 1897 in Nigeria, where he was Senior Medical Officer on Lugard's staff.

After a year as Superintendent of the LSTM, however, he decided (in 1901) to emigrate to South Africa, where he became District Surgeon and Port Health Officer at Port Elizabeth, Natal. He went to South Africa initially as a physician in the plague service (he was adept at controlling epidemics of plague, smallpox and typhus), and subsequently acted in various capacities in the Health Department. Rees himself died of typhus at the age of 49 years, and a memorial tablet was erected in Port Elizabeth Hospital. Rees also contributed letters on beri-beri and malaria. Rees's son (Harland) became a consultant surgeon in London.[5]

Charles Wilberforce Daniels (1862–1927)

Charles Wilberforce Daniels (1862–1927; Figure 19.2) was born in Manchester, and at thirteen years of age entered Manchester Grammar School; four years later he won a scholarship to Trinity College, Cambridge. His medical education was obtained at the (Royal) London Hospital, and after graduating in 1886 he served there as house surgeon and house physician. Daniels joined the Colonial Medical Service in 1887, and served in Fiji, British Guiana and the West Indies. In 1898 he travelled to Calcutta on behalf of the Royal Society to confirm Ross's malarial researches, and the following year he went to Nyasaland (now Malawi).

Daniels returned to England in 1900, and the following year was appointed second Superintendent of the LSTM. At Manson's request, however, he left London temporarily to become Director of the newly-founded Institute of Medical Research at Kuala Lumpur. On relinquishing this appointment after two years he again became Superintendent of the LSTM, but later resigned from this post to become Lecturer in Tropical Medicine. Daniels was also a lecturer at the (Royal) London and St George's Hospitals, and the London School of Medicine for Women (later the Royal Free School of Medicine). When Manson retired he became Medical Officer to the Colonial Office; in 1919 he resigned due to a progressive neurological illness.[6]

George Carmichael Low (1872–1952)

Details of the life and career of George Carmichael Low have been outlined in Chapter 8. He became the third Superintendent of the LSTM in succession to Daniels, who in turn succeeded him in this capacity in 1905.[7]

Hugh Basil Greaves Newham (1874–1959)

Hugh Basil Greaves Newham was born at Winslow, Buckinghamshire, and educated at Wellingborough and Malvern Schools, and St Thomas's Hospital, qualifying

FIGURE 19.2 Charles Wilberforce Daniels (1862–1927), second Superintendent of the London School of Tropical Medicine (reproduced courtesy of The Wellcome Library, London).

in 1898. For the next two years, during which time his interest in tropical medicine was aroused, he served as a ship's surgeon. Having obtained the DPH in 1900, he spent six years in general practice in Liverpool.

Newham moved to London in 1906 to join the LSTM. He obtained the MD (Durham) in 1917. He succeeded Daniels (now designated Director of the LSTM) in 1910. According to Manson-Bahr, his lectures were so good that the laboratory at the ADH had to be enlarged in 1912 to accommodate 65 students, together with their living quarters. He continued as Director into the Great War (1914–18), but in 1917 he became Consultant Physician in Tropical Diseases to the East African Forces, and served in Kenya, Uganda and Tanganyika (now Tanzania).

When he returned to London, the LSTM had removed (and was now entitled the HTD) to Euston; there, he resumed activities as Director and Superintendent, and in 1929 transferred to the newly opened London School of Hygiene and Tropical Medicine (LSHTM). He retired from the LSHTM in 1938; during his time there he organized (having performed a similar function at Endsleigh Gardens) the Museum, which was partly destroyed by enemy action in 1941. Newham also published several papers in medical journals (*see Medical Directory*).[8]

William Ernest Cooke (1879–1967)

Cooke, an Irishman, who was educated at Wesley College, Dublin, and the Royal College of Physicians and Surgeons, Dublin, succeeded Newham as Superintendent of the HTD from 1925 until 1935, but is probably best remembered for his contributions to *clinical* Tropical Medicine during the Second World War (1939–45). During these years, following the condemnation of the Endsleigh Gardens building in 1939, he held regular out-patient sessions in a room at the LSHTM.

Cooke organized the out-patient department of the HTD so successfully that by 1935 the annual attendance exceeded 2000.[9]

Sir Perceval Alleyn Nairne (1841–1921)

Perceval Alleyn Nairne was the fifth son of Captain Alexander Nairne, RN and HEICS, who himself was a member of the Committee of Management of the Seamen's Hospital Society for 30 years, beginning in 1886. Nairne became Chairman of the SHS from 1898 until 1921, and also Chairman of the LSTM from 1899. He was a solicitor by training, and was also a prominent freemason. He had a long association with seafaring folk, and was a lover of the sea, being a keen yachtsman and owning his own schooner for some twenty years. He was of very conservative taste, and apparently 'hated change of any kind'.

Nairne was, above all, a staunch advocate for the establishment of the LSTM, and therefore of this new discipline.[10]

Sir Pietro James Michelli (1853–1935)

Pietro James Michelli (1853–1935), an Irishman, was Secretary of the newly-established LSTM from 1899 until 1924, when he handed over to the Governing Body of the newly-founded LSHTM. Michelli was described by Manson-Bahr as the 'mainstay of the whole institution' and a great 'fundraiser'. Manson-Bahr ends his sketch: 'It was he who held the financial tiller and so guided the infant [LSTM] into a safe harbour and ... nourished and cared for it in its growing years'.[11]

NOTES

1 G C Low. A retrospect of Tropical Medicine from 1894 to 1914. *Trans R Soc Trop Med Hyg* 1929, 23: 213–34; R S Desowitz. *Tropical Diseases from 50,000 BC to 2500 AD*. London, 1997: HarperCollins, pp. 145–7; P Manson-Bahr. Francis Lovell, Havelock Charles. In: *History of the*

School of Tropical Medicine in London (1899–1949). London, 1956: H K Lewis, pp. 174–6; Anonymous. Sir Havelock Charles Bt. *Br Med J* 1934, 29 November.

2 Anonymous. *Liverpool School of Tropical Medicine: historical record 1898–1920*. Liverpool, 1920: The University Press, p. 103; P J Miller. *'Malaria, Liverpool': an illustrated history of the Liverpool School of Tropical Medicine 1898–1998*. Liverpool, 1998: Liverpool School of Tropical Medicine, p. 78.

3 *Ibid*. (Anonymous).

4 Anonymous. *Times, Lond* 1911, 19 June; Anonymous. *Lancet* 1911, i: 59–60; Anonymous. *Br Med J* 1911, i: 53–4; C S Sherrington, C E J Herrick. Boyce, Sir Rubert William (1863–1911). In: H C G Matthew, B Harrison (eds), *Oxford Dictionary of National Biography*, Vol. 7. Oxford, 2004: Oxford University Press, pp. 15–16.

5 Anonymous. David Charles Rees. *Lancet* 1917, ii: 549; Anonymous. David Charles Rees. *Br Med J* 1917, ii: 469–70; Anonymous. *Medical Directory*. London, 1900: J & A Churchill, p. 331; P Manson-Bahr. Dr David Charles Rees. In: *History of the School of Tropical Medicine in London (1899–1949)*. London, 1956: H K Lewis, p. 158. See also: D C Rees. Malarial crescents and spheres. *Br Med J* 1898, i: 491; Anonymous. Harland Rees. *Times, Lond* 2002.

6 Anonymous. *Medical Directory*. London, 1924: J & A Churchill, p. 560; P Manson-Bahr. Dr Charles Wilberforce Daniels. In: *A History of the School of Tropical Medicine in London (1899–1949)*. London, 1956: H K Lewis, pp. 162–5; G C Cook. Charles Wilberforce Daniels FRCP (1862–1927): underrated pioneer of tropical medicine. *Acta Tropica* 2002, 81: 237–50.

7 *Op cit*. See note 1 above (Manson-Bahr), pp. 158–62; G C Cook. George Carmichael Low FRCP: an underrated figure in British tropical medicine. *J R Coll Phys Lond* 1993, 27: 81–2; G C Cook. George Carmichael Low FRCP: twelfth president of the Society and underrated pioneer of tropical medicine. *Trans R Soc Trop Med Hyg* 1993, 87: 355–60; M Worboys. Low, George Carmichael (1872–1952). In: *Oxford Dictionary of National Biography*, Vol. 34. Oxford, 2004: Oxford University Press, pp. 550–51.

8 Anonymous. *Medical Directory*. London, 1930: J & A Churchill, p. 239; P Manson-Bahr. In: *History of the School of Tropical Medicine in London (1899–1949)*. London, 1956: H K Lewis, pp. 165–6; Anonymous. Hugh Basil Greaves Newham. *Lancet* 1959, ii: 978, 1096; Anonymous. H B G Newham. *Br Med J* 1959, ii: 1336–7.

9 Anonymous. *Medical Directory*. London, 1960: J & A Churchill, p. 436; P Manson-Bahr. In: *A History of the School of Tropical Medicine in London (1899–1949)*. London, 1956: H K Lewis, pp. 147–8; G C Cook. William Ernest Cooke FRCSI (1879–1967) and his Asian tour of 1929–1930. *Acta Tropica* 2006, 100: 1–10.

10 Anonymous. *Times, Lond* 1921, 11 December; Anonymous. Nairns, Sir Perceval Alleyn. *Who Was Who 1916–1928*. London, 1947: A & C Black, p. 593; P Manson-Bahr. Sir Perceval Alleyn Nairne. In: *History of the School of Tropical Medicine in London (1899–1949)*. London, 1956: H K Lewis, pp. 125–6.

11 Anonymous. *Times, Lond* 1935, April 13; Anonymous. Michelli, Sir James. *Who Was Who 1929–1940*. London, 1941: A & C Black, p. 937; P Manson-Bahr. Sir Pietro James Michelli. In: *History of the School of Tropical Medicine in London (1899–1949)*. London, 1956: H K Lewis, pp. 126–7.

20

Politicians and entrepreneurs: the Chamberlains (father and son), Alfred Jones and Herbert Read

It is impossible to consider the pioneers of the formal discipline without mention of some of those who were responsible, both politically and financially, for the two pioneering schools, situated in Liverpool and London, respectively. Without the impetus from these men, the discipline could not possibly have been founded. Therefore this final chapter focuses on the Chamberlains (father and son), and others who were the 'driving forces' behind the London and Liverpool Schools of Tropical Medicine, with financial assistance from HM Government, The Royal Society and private funds. The shipping magnate, Sir Alfred Jones, was both an enthusiast and donor of funds for the foundation of the Liverpool School of Tropical Medicine. However, the major inspiration for this School came from Sir Rubert Boyce (see Chapter 19), who was the first Dean; although a distinguished medical graduate, his role here was largely administrative.

JOSEPH CHAMBERLAIN MP, FRS (1836–1914)

Chamberlain (Figure 20.1) was to become a leading parliamentarian as Colonial Secretary, and in many ways the prime instigator of the new discipline of tropical medicine.

Tropical Medicine: An Illustrated History of The Pioneers

GOING TOO FAR.

Right Hon. J-s-ph Ch-mb-rl-n (in his Orchid-house). "RHODES MAY SAY WHAT HE LIKES ABOUT 'UNCTUOUS RECTITUDE, BUT WHEN HE SPEAKS DISRESPECTFULLY OF MY ORCHID——!!"

[" You know every man must do something. Some people grow orchids."—*Extract from Mr. Cecil Rhodes' Speech at the Guildhall, Capetown.*]

FIGURE 20.1   *Punch* cartoon (1897, January 16: 26) depicting Joseph Chamberlain as an ardent orchid grower. As well as having a monocle, Chamberlain always stood out from other politicians because of his 'button-hole' orchid.

Joe Chamberlain was born at Camberwell, Surrey, the first of nine children of Joseph Chamberlain (1796–1874), a cordwainer and metal manufacturer. He was a Unitarian – unlike most Conservative MPs, who were at that time staunch members of the established church – and an industrialist. Four generations of his family had worked near Guildhall, London, as cordwainers or shoe manufacturers, the demand for which obviously escalated in wartime. Chamberlain's grandfather was a dedicated supporter of the Honourable Artillery Company, and became Captain of their north-western division. There was thus an imperial dimension to his background. Chamberlain derived a literary background from his mother, Caroline Harben's (1808–75) family; her father was a brewer and provisions merchant, and the family enjoyed good food and drink, and delighted in plays.

Following education at a dame-school in Camberwell, which was conducted by a Church of England clergyman, and University College School, London – much favoured by Unitarians – he was apprenticed to his father's business, during which time he travelled to Belgium and France as translator for his cordwainer uncle. Joseph's father then invested in a wood-screw business (which belonged to his brother-in-law) in Birmingham; he worked there for some fifteen years, and became one of the 'Screw Kings' of that city, taking charge of the accounting side of the business. He marketed goods so successfully that producers of screws in the USA paid the firm a large annual fee to stay away! This convinced him of the compatibility of imperialism and free trade. He also joined the board of directors of the Midland Railway.

Chamberlain later jointly founded the National Education League, which sought parliamentary legislation to create a network of schools, under civic control and fully funded by taxation, to instil into every child basic literacy and numeracy. It was this that propelled him into national politics (see below).

His first marriage (see below) produced a daughter and a son (Joseph) Austen, and his second a son (Arthur) Neville (1869–1940) who became British Prime Minister, and three daughters.

Political career

Chamberlain's political career began in 1873 when he was first elected Mayor of Birmingham, and in June 1876 he entered the House of Commons in Gladstone's Government as a radical liberal. By the time he became Mayor, Chamberlain was a very wealthy man who devoted much time and energy not to business interests but to the Birmingham civic arena. He built himself a substantial house in Birmingham – 'Highbury' – and there became an avid enthusiast of orchid-growing.

In 1877, Chamberlain established the National Federation of Liberal Associations, which had Gladstone's support. He entered the Cabinet as President of the Board of Trade at a time (in the 1880s) when upheavals in Ireland and

Africa dominated the scenario. An attack on ship-owners for acts of negligence – which had caused significant mortality in the merchant navy – failed to endear him widely, although he assumed a prominent position on imperial matters and was greatly involved with South Africa.

The 'home-rule' crisis, in which Chamberlain was involved, deprived the Liberal Party power for twenty years. When the liberals fell from office in 1895, the third Marquess of Salisbury (Prime Minister) called Chamberlain and the Duke of Devonshire into council to decide on the terms for and composition of a coalition Unionist ministry. Chamberlain chose the post of Secretary of State for the Colonies, thus 'gratifying [his] sense of *imperial* mission'. Salisbury, meanwhile, added the post of Foreign Secretary to that of Prime Minister.

Chamberlain and tropical medicine

P T Marsh has summarized Joseph Chamberlain's monumental contributions to tropical medicine between 1895 and 1903:[1]

> [He] made arguably his most substantial contribution to the crown colonies by promoting research and education in the field of *tropical medicine* [my italics]. Concern for education permeated Chamberlain's career in public life from its beginnings with the National Education League to the establishment of the University of Birmingham at the turn of the century. As colonial secretary he looked to education to address a problem that crippled *imperial* [my italics] administration especially in tropical west Africa, where most men sent out from England were stricken with malaria before they could complete their assignments. He appointed ... Manson, the best known of the few experts in tropical medicine, to be the Colonial Office medical adviser.
>
> Chamberlain raised private as well as public money to found the *London School of Tropical Medicine* [my italics]. He fostered the appointment of committees of British researchers to go to the tropics not only for on-site investigation but to make and execute policy. From the sanitary reforms he made as mayor of Birmingham to the transformation he brought about in the sanitation of the camps into which Boer women and children were herded during the second South African War, even his opponents recognized that he knew how to deal effectively with public health. Before he left office he had made service in the tropical colonies a more attractive profession.
>
> One way and another, Chamberlain's administrative efforts as colonial secretary marked him out as the most outstanding holder of that formerly secondary office, while the importance of imperial issues at the turn of the century and his personal importance in the party politics of the day lifted his position to an eminence second only to the prime minister. Even his rivals treated him as the first minister of the empire.

The South African war terminated in 1902; this was followed by Salisbury's retirement, leaving the government an uneasy dual leadership with Balfour as Prime Minister and Chamberlain as 'first minister of the empire'. In September 1903, Chamberlain resigned from the Colonial Office. Three years later he suffered a stroke, which terminated his political career, but he survived for another seven years; he had retired from the House of Commons when he died on 2 July 1914.

A burial at Westminster Abbey was declined, at his request, and he was interred in West Birmingham on 6 July.[2]

(JOSEPH) AUSTEN CHAMBERLAIN KG, MP (1863–1937)

Two years after his first marriage, which had already produced a daughter, Joseph Chamberlain's wife died in childbirth, having delivered (Joseph) Austen (Figure 20.2). Austen was educated at Rugby and Trinity College, Cambridge. Following nine months in France and a further twelve months in Berlin, in March 1892 he was elected MP for East Worcestershire as a liberal unionist. When the Conservative Party was returned to office in 1895 he was made a civil Lord of the Admiralty, and in 1900 he became Financial Secretary to the Treasury. This led to a position in the Cabinet, and in 1903 he was appointed Chancellor of the Exchequer. He apparently had a somewhat difficult time with the Prime Minister, Balfour, and, following his resignation, also with Bonar Law. In 1914, the year in which his father died, he succeeded to the constituency of West Birmingham.

In May 1915 he became Secretary of State for India for two years, and in April 1918 he became a member of the War Cabinet. When Lloyd George reorganized his government after the general election of 1918, Austen Chamberlain again became Chancellor of the Exchequer. When Bonar Law retired on account of ill health (he had an inoperable malignant condition) in March 1921, Chamberlain succeeded him as Conservative leader; problems associated with Ireland and Turkey loomed large during this period. Defeated on a policy matter, he ceased to be leader of the Conservative Party. In November 1924, Baldwin, in his second administration, offered him the Foreign Secretaryship – the post for which he is best remembered – which was gladly accepted. He was appointed a Knight of the Garter, and regularly attended meetings of the League of Nations. At the General Election of 1929, the Conservative Government was defeated and the administration resigned.

When an all-party government was formed in August 1931, he was disappointed that he was offered the Admiralty rather than the Foreign Office; after the general election of October of that year, Austen Chamberlain wrote to Baldwin indicating that he did not want any part in his new administration. In 1932 he served on a joint select committee on Indian constitutional reform, and also took a role in several non-political activities – the Chancellorship of Reading University (1935–7), the Chairmanship of the Court of Governors of the London School of Hygiene and Tropical Medicine (LSHTM) and of the governing body of the British Postgraduate Medical School; he was also Chairman of the Board of Governors of Rugby School. Chamberlain died suddenly on 16 March 1937.[3]

ALFRED JONES (1845–1909)

Alfred Jones is described in the *Oxford Dictionary of National Biography* as a 'shipping entrepreneur and colonial magnate'. Jones was born in Carmarthen, and moved to Liverpool with his parents at the age of three years. Educated at

IN STATUE QUO.

Rt. Hon. J-s-ph Ch-mb-rl-n (on his travels, after consulting Guide-book). "' THE EMPEROR CALIGULA MADE HIS HORSE A CONSUL.' LET ME SEE, AUSTEN, WHAT DID I DO FOR JESSE COLLINS?"

FIGURE 20.2 *Punch* cartoon (1900, November 18: 383) showing Austen with his father, Joseph – both of whom wore monocles, and were therefore both 'easy prey' for cartoonists.

local schools, he started an association with shipping in 1859 as a cabin-boy with the African Steam-Ship Company. He then attended evening classes at Liverpool College, and subsequently became manager of McGregor Laird. In 1878, he started his own firm as a shipping and insurance broker. He was later made junior partner in the Elder Dempster Line Ltd; in 1884 the senior partners were 'bought out' by Jones, and by 1909 (at the time of his death) the firm – which concentrated on West Africa – possessed no less than 101 ships.

Throughout his life, Jones apparently held West Africans in high regard, and his primary motivation was to assist them whenever he could. In 1898, a financial contribution led to the formation of the Liverpool School of Tropical Medicine – largely to deal with the high levels of mortality in both the indigenous and expatriate populations of West Africa.

His philanthropic activities were broad, however, and involved missionary societies, Liverpool University, the Anglican cathedral, the Liverpool Training League, and Lancashire Sea Training Home. Jones was knighted in 1901 for these charitable works.[4] He was later open to criticism for assumed connections with Congo brutalities as a result of a 'special relationship' with King Leopold II.[5]

HERBERT JAMES READ GCMG (1863–1949)

Herbert Read was born at Honiton, Devon, and educated at Allhallows School, Honiton, and Brasenose College, Oxford. He entered the War Office as a higher division clerk in 1887, and after two years was transferred to the Colonial Office, where he stayed for over 40 years. At various times he was Head of the West and East African Departments and later Assistant Under-Secretary, with supervision of both departments.

Read, a close friend of Manson, was Joseph Chamberlain's assistant private secretary from 1896–7; it was probably he who instigated Manson's appointment as Medical Adviser to the Colonial Office in 1897. He was a civil servant who was devoted to the Colonial Service. He had also been Governor of Mauritius (1924–30) and, following retirement, was on the governing bodies of the Seamen's Hospital Society (SHS) and LSHTM. Africa (especially tropical Africa) was thus very close to his heart.

Manson could not possibly have founded the London School of Tropical Medicine (see Chapter 3) alone; he required assistance from laymen of the SHS, and from numerous able administrators, such as Sir Herbert Read.[6]

ESTABLISHMENT OF THE PIONEERING SCHOOLS OF TROPICAL MEDICINE IN BRITAIN

The Chamberlains, father and son, both eminent politicians, thus made substantial contributions to the new discipline. Both London Schools were subsequently recognized for their very different contributions, one curative and the other essentially preventive – the LSTM in the case of Joseph, and the LSHTM in Austen's case.

Low has outlined the early days of the Liverpool School of Tropical Medicine, which was opened some six months before the School in London – and must therefore be designated as *the* pioneering School. On 20 January 1899 a Dean for this School was appointed, and on 7 February a Demonstrator in Tropical Pathology (Annett) followed. On 10 April a Lecturer in Tropical Medicine was appointed, and teaching started the following month. On 22 April 1899, the School was officially opened. In 1902 J W W Stephens followed Annett, and in 1909 Warrington Yorke was appointed Research Assistant at the Runcorn Laboratory, becoming Director the following year. Stephens was promoted to the Alfred Jones Chair of Tropical Medicine in 1913, and in the same year Yorke was appointed to the Walter Myers Chair of Parasitology (Myers had died of yellow fever – see Chapter 6 – while researching the disease in 1901). By 1913, the School had sent no less than 31 expeditions to the tropics.

In his Presidential Address to the Royal Society of Tropical Medicine and Hygiene in 1929, Low in addition gave a succinct account of the foundation of the *London* School of Tropical Medicine (see also Chapter 3). Manson had begun lecturing to London audiences on tropical diseases in 1894, and he was increasingly aware that, outside the Army and IMS, medical practitioners 'should have special teaching [in this specialty] before proceeding abroad'. In July 1897 he was appointed Medical Adviser to the Colonial Office on the advice of Read (see above), and he came 'into intimate touch' with (Joseph) Chamberlain, described as 'one of the most brilliant and far-seeing statesmen in England at that time! On 11 March 1898 Chamberlain addressed a circular letter to the GMC and leading medical schools of the UK, pointing out

> the importance of ensuring that all medical officers selected for appointments in the tropics should enter on their careers with expert knowledge requisite for dealing with such diseases as are prevalent in tropical climates, [and therefore medical officers of the Colonies] should be given facilities in the various medical schools for obtaining some preliminary knowledge of the subject.

The General Medical Council replied favourably, and on 28 May of that year Chamberlain addressed a circular to the Governors of all the Colonies. Manson had by then succeeded Sir Charles Gage Brown, and his attention was 'directed to the importance of scientific enquiry into the causes of malaria, and of special education in *tropical medicine* [my italics] for the medical officers of the Crown Colonies'.

The London School started teaching courses on 2 October 1899, the staff having been appointed in May of that year. Apparently, as early as May 1897, Manson had begun negotiations with Michelli (Secretary of the SHS; see Chapter 19) for the establishment of his School at the Albert Dock Hospital; the Board of the SHS 'acceded to [Chamberlain's] request', and on 14 October 1898 a sub-committee (with Read [see above] representing the Colonial Office) was appointed to organize and manage 'the new School of Tropical Medicine with powers to co-opt members of the medical profession interested in tropical diseases and bacteriology'. After several meetings, the rules and suitable advertisements for members of the teaching staff, were drawn up on 10 March

1899. In addition to Manson himself, Cantlie, Simpson, Duncan, Baker, Hewlett and Sambon were appointed to the staff. The first Medical tutor was Rees (see Chapter 19). Leiper (a helminthologist) joined the School in January 1905, and Wenyon (a protozoologist) the following May. The department of entomology, with Alcock in charge, was added later.[7]

These initiatives resulted in the establishment of the two pioneering Schools of *Tropical Medicine*.

NOTES

1 P T Marsh. Chamberlain, Joseph (Joe) (1836–1914). In: H C G Matthew, B Harrison (eds), *Oxford Dictionary of National Biography*, Vol. 10. Oxford, 2004: Oxford University Press, pp. 923–34.

2 *Ibid*; R Jay. *Joseph Chamberlain: a political study*. Oxford, 1981: Clarendon Press, p. 383; P T Marsh. *Joseph Chamberlain, Entrepreneur in Politics*. London, 1994: Yale University Press, p. 724. See also G M Thomson. *The Prime Ministers from Robert Walpole to Margaret Thatcher*. London, 1980: Secker & Warburg, p. 260.

3 Anonymous. *Times, Lond* 1937, 17 March; D J Dutton. Chamberlain, Sir (Joseph) Austen (1863–1937). In: H C G Matthew, B Harrison (eds), *Oxford Dictionary of National Biography*, Vol. 10. Oxford, 2004: Oxford University Press, pp. 906–14. See also G C Cook. *John MacAlister's Other Vision: history of the Fellowship of Postgraduate Medicine*. Oxford, 2005: Radcliffe Publishing, p. 181.

4 P N Davies. *Sir Alfred Jones: shipping entrepreneur par excellence*. London, 1978: Europa Publications Ltd, p. 162; J G Read. Jones, Sir Alfred Lewis (1845–1909). In: H C G Matthew, B Harrison (eds), *Oxford Dictionary of National Biography*, Vol. 30. Oxford, 2004: Oxford University Press, pp. 436–8.

5 J Conrad. *Heart of Darkness*. London, 1994: Penguin Books, p. 111; A Hochschild. *King Leopold's Ghost: a story of greed, terror, and heroism in Colonial Africa*. London, 1999: MacMillan, p. 366.

6 Anonymous. Sir Herbert Read: medicine and colonial administration. *Times, Lond* 1949, October 18; P Manson-Bahr. Sir Herbert Read. In: *History of the School of Tropical Medicine in London (1899–1949)*. London, 1956: H K Lewis, pp. 127–9; Anonymous. Read, Sir Herbert James. *Who Was Who 1941–1950*, Vol. 4. London, 1951: A & C Black, p. 959; E W Evans, A Jackson. Read, Sir Herbert James (1863–1949). In: H C G Matthew, B Harrison (eds), *Oxford Dictionary of National Biography*, Vol. 46. Oxford, 2004: Oxford University Press, pp. 212–13.

7 G C Low. A retrospect of Tropical Medicine from 1894 to 1914. *Trans R Soc Trop Med Hyg* 1929, 23: 213–32; Anonymous. *Liverpool School of Tropical Medicine: historical record 1898–1920*. Liverpool, 1920: University Press of Liverpool, p. 103; P Manson-Bahr. *History of the School of Tropical Medicine in London (1899–1949)*. London, 1956: H K Lewis, p. 328; P J Miller. '*Malaria Liverpool': an illustrated history of the Liverpool School of Tropical Medicine 1898–1998*. Liverpool, 1998: Liverpool School of Tropical Medicine, p. 78.

Epilogue

'Medicine in the tropics' will always be a viable pursuit; however, variations will arise as '*new*' entities (such as HIV/AIDS) evolve, and as increasing urbanization takes place. However, the formal discipline, which arose in the latter days of Victoria's long reign, has been on the decline for several decades.

This downward trend has coincided with the loss of Britain's Empire and colonies, and also that of former colonial powers.[1] It has in addition much to do with several other factors. Diagnosis has become far simpler, and the majority of parasitic infections (which formed the backbone of the formal discipline) can be either confirmed or refuted by serological techniques.[2] Treatment of these is now both more effective and safer than in former days; modern prophylaxis and chemotherapy have revolutionized the scenario. The discipline has thus forfeited most of its former mystique! This has inevitably led to management of what were formerly considered 'exotic' diseases in infectious disease units, without referral to a tropical centre – such as those previously located in Liverpool or London.

Now, therefore, the pioneers of the formal discipline have been relegated to medical history, and their discoveries, many of them outlined in this book, have been applied to 'medicine in the tropics' – a discipline *without* colonial interests.

NOTES

1 J Morris. *Farewell the Trumpets: an Imperial retreat*. London, 1992: The Folio Society, p. 482; L James. *The Rise and Fall of the British Empire*. London, 1994: Little Brown and Co, p. 704.

2 D I Grove. *A History of Human Helminthology*. Wallingford, 1990: CAB International; G C Cook. History of parasitology. In: S H Gillespie, R D Pearson (eds), *Principles and Practice of Clinical Parasitology*. Chichester, 2001: John Wiley & Sons, pp. 1–20; F E G Cox. History of human parasitic diseases. *Infect Dis Clin North Am* 2004, 18: 171–88.

Appendices

Appendix I: Presidential Addresses to the (Royal) Society of Tropical Medicine and Hygiene devoted to the history of tropical medicine and hygiene

Date of address	President	Reference[a]	Subject
1907[b]	P Manson	1, 1–12	Survey of tropical medicine to 1907
1909[b]	R Ross	2, 272–88	History of infectious disease and sanitation
1919[b]	W J R Simpson	13, 31–44	Preventable diseases
1925	A Balfour	19, 189–231	Pioneers of medicine in the tropics
1929	G C Low	23, 213–34	Tropical medicine from 1894 to 1914
1933	L Rogers	27, 217–38	Cholera, smallpox and plague
1935[b]	A G Bagshawe	29, 211–26	Disease in some small tropical islands
1943	H H Scott	37, 169–88	Slave-trade and the spread of tropical disease
1951[b]	N H Fairley	45, 279–303	Schistosomiasis
1959	W MacArthur	53, 423–39	Pestilences of the past
1961	G McRobert	55, 485–96	Transition from Empire to Commonwealth
1963	C Wilcocks	57, 395–408	History of preventive, social, and occupational medicine
1967	P C C Garnham	61, 753–64	Early researches into malaria
1983[b]	I McGregor	78, 1–8	Malaria research
1989[b]	G S Nelson	84, 3–13	Filariasis, trichinosis, hydatid disease, schistosomiasis
1993	G C Cook	88, 4–18	Origin(s) of tropical medicine & the (Royal) STMH

[a] References are given as volume and page numbers in the *Transactions of the (Royal) Society of Tropical Medicine and Hygiene*.
[b] Only part of the Address devoted to this topic.

Tropical Medicine: An Illustrated History of The Pioneers
Copyright © 2007 Elsevier Ltd. All rights of reproduction in any form reserved.

Appendix II: Some major 'milestones' in the history of tropical medicine

Date (year)	Event
1550 BC	Visualization of *Dracunculus medinensis* (the Guinea worm).[a]
600 BC	Association of 'the fever' with mosquitoes.[b]
1803	Publication of the Thomas Winterbottom's (1766–1859) book, *An account of the native Africans in the neighbourhood of Sierra Leone, to which is added an account of the present state of medicine among them*, 2nd edn. London, 1969: F Cass & Co Ltd, pp. 362, 283.
1821	Foundation of the Seamen's Hospital Society [SHS] (8 March).
1832	Publication of William Twining's (1790–1835) book, *Clinical illustrations of the more important diseases of Bengal, with the result of an enquiry into their pathology and treatment*. Calcutta, 1832: Baptist Mission Press, p. 707.[c]
1844	Patrick Manson born (3 October).
1845	Alphonse Laveran born (18 June).
1855	David Bruce born (29 May).
1857	Ronald Ross born (13 May).
1877	Birth of the formal discipline. Manson's demonstration of mosquito involvement in lymphatic filariasis (10 August), and letter to T S Cobbold (27 November).
1880	Laveran's demonstration of the malaria parasite (6 November).
1897	Ross's demonstration of mosquito involvement in malaria transmission (20 August).
1898	Ross's demonstration of the life-cycle of *Proteosoma* spp (avian malaria); Manson's textbook devoted to disease in warm climates published.
1899	Foundation of Liverpool School of Tropical Medicine (22 April); foundation of London School of Tropical Medicine (at Albert Dock [SHS] Hospital) (2 October).
1900	George Low establishes mosquito-to-man transmission of lymphatic filariasis; Schools of Tropical Medicine founded in Paris, Hamburg, Antwerp.
1901	Role of mosquito in transmission of yellow fever established by American Yellow Fever Commission.
1907	Society of Tropical Medicine and Hygiene (STMH) founded.
1912	Japanese workers establish life-cycle of *Schistosoma japonicum*.
1920	LSTM moves to Endsleigh Gardens WC1; HTD founded on 11 November (under SHS auspices until 1948); STMH becomes the *Royal* Society of Tropical Medicine & Hygiene; Calcutta School of Tropical Medicine opened.
1922	Manson dies (9 April); Laveran dies (18 May).
1926	Ross Institute & HTD founded (RIHTD) (15 July).
1929	London School of Hygiene and Tropical Medicine (LSHTM) opened as Rockefeller-funded Public Health School (15 July).
1931	Bruce dies (27 November).
1932	Ross dies (16 September).

Date (year)	Event
1934	RIHTD moves to central London – *clinical* Tropical Medicine to HTD; basic sciences to LSHTM.
1946	Neil Fairley becomes first occupant of Wellcome chair of Clinical Tropical Medicine in London; Discussion of 'Imperial' Hospital in Bloomsbury; SHS abandons HTD to NHS – NHS Act passed.
1948	NHS founded (5 July).
1951	HTD removes from Devonshire Street to St Pancras (24 May).
1960 onwards	Insidious decline in the 'formal discipline'; 'medicine in the tropics' becomes increasingly important, as does travel medicine.

[a] Papyrus Ebers (*c.* 1550 BC); Holy Bible, Numbers 21: 6; G C Cook. Discovery and clinical importance of the filariases. *Infect Dis Clin North Am* 2004, 18: 219–30.

[b] G C Cook. *From the Greenwich Hulks to Old St Pancras: a history of tropical disease in London.* London, 1992: Athlone Press, p. 338; C M Poser, G W Bruyn. In: *An Illustrated History of Malaria.* London, 1999: Parthenon Publishing Group, pp. 121–5.

[c] G C Cook. William Twining (1790–1835): the first accurate descriptions of 'tropical sprue' and kala-azar? *J Med Biog* 2001, 9: 125–31.

INDEX